ZEW Economic Studies

Publication Series of the Centre for European Economic Research (ZEW), Mannheim, Germany

ZEW Economic Studies

Further volumes of this series can be found at our homepage:
http://www.springer.de/cgi/svcat/search_book.pl?series=4041

Vol. 6: B. Fitzenberger
Wages and Employment Across Skill Groups
An Analysis for West Germany
1999. XII, 251 pp. ISBN 3-7908-1235-8

Vol. 7: K. Rennings et al. (Eds.)
Social Costs and Sustainable Mobility
Strategies and Experiences in Europe
and the United States
1999. VI, 212 pp. ISBN 3-7908-1260-9

Vol. 8: H. Legler et al.
Germany's Technological Performance
2000. X, 191 pp. ISBN 3-7908-1281-1

Vol. 9: O. Bürgel
The Internationalization of British Start-up Companies in High-Technology Industries
2000. XIV, 230 pp. ISBN 3-7908-1292-7

Vol. 10: J. Hemmelskamp et al. (Eds.)
Innovation-Oriented Environmental Regulation
Theoretical Approaches and Empirical Analysis
2000. VI, 348 pp. ISBN 3-7908-1313-3

Vol. 11: K.L. Brockmann, M. Stronzik (Eds.)
Flexible Mechanisms for an Efficient Climate Policy
Cost Saving Policies and Business Opportunities
2000. VIII, 130 pp. ISBN 3-7908-1314-1

Vol. 12: W. Smolny
Endogenous Innovations and Knowledge Spillovers
A Theoretical and Empirical Analysis
2000. VIII, 277 pp. ISBN 3-7908-1320-6

Vol. 13: M. Lechner, F. Pfeiffer (Eds.)
Econometric Evaluation of Labour Market Policies
2001. X, 242 pp. ISBN 3-7908-1372-9

Vol. 14: M. Beise
Lead Markets
Country-Specific Success Factors of the Global Diffusion of Innovations
2001. XVI, 305 pp. ISBN 3-7908-1430-X

Vol. 15: O.H. Jacobs, C. Spengel
Effective Tax Burden in Europe
Current Situation, Past Developments and Simulations of Reforms
2002. X, 131 pp. ISBN 3-7908-1470-9

Vol. 16: U. Kaiser
Innovation, Employment, and Firm Performance in the German Service Sector
2002. X, 164 pp. ISBN 3-7908-1481-4

Vol. 17: J. Köke
Corporate Governance in Germany
An Empirical Investigation
2002. VIII, 184 pp. ISBN 3-7908-1511-X

Vol. 18: E. Wolf
What Hampers Part-Time Work?
An Empirical Analysis of Wages, Hours Restrictions and Employment from a Dutch-German Perspective
2003. X, 174 pp. ISBN 3-7908-0006-6

Vol. 19: P. Cecchini et al. (Eds.)
The Incomplete European Market for Financial Services
2003. XII, 255 pp. ISBN 3-7908-0013-9

Vol. 20: C. Böhringer, A. Löschel (Eds.)
Empirical Modeling of the Economy and the Environment
2003. VI, 326 pp. ISBN 3-7908-0078-3

Klaus Rennings · Thomas Zwick (Eds.)

Employment Impacts of Cleaner Production

With 100 Figures
and 50 Tables

Physica-Verlag
A Springer-Verlag Company

ZEW
Zentrum für Europäische
Wirtschaftsforschung GmbH
Centre for European
Economic Research

Series Editor
Prof. Dr. Wolfgang Franz

Editors

Dr. Klaus Rennings
Dr. Thomas Zwick
Centre for European Economic Research (ZEW)
L7, 1
68161 Mannheim, Germany

ISBN 3-7908-0093-7 Physica-Verlag Heidelberg New York

Cataloging-in-Publication Data applied for
Bibliographic information published by Die Deutsche Bibliothek
Die Deutsche Bibliothek lists this publication in the Deutsche Nationalbibliografie; detailed bibliographic data is available in the Internet at <http://dnb.ddb.de>.

Physica-Verlag Heidelberg New York
a member of BertelsmannSpringer Science+Business Media GmbH

Cover design: Erich Dichiser, ZEW, Mannheim

SPIN 10940730 88/3130-5 4 3 2 1 0 – Printed on acid-free paper

Preface

This book covers a highly relevant but clearly under-researched topic – the impact of environmental innovations on employment on the firm level.

While the impact of innovations on employment has been analysed empirically in several studies, only few studies have specifically analysed innovations which are environmentally beneficial. In this book, we present case studies and an in-depth analysis of the behaviour of more than 1,500 firms that have recently introduced environmental innovations in five European countries (Germany, United Kingdom, Italy, the Netherlands, Switzerland). The data stem from identical telephone interviews carried out simultaneously in these countries. The interviews were especially designed for analysing the relationship between environmental innovation and employment on the firm-level. On the basis of an econometric model we find which factors lead to positive and which factors lead to negative employment effects in the wake of environmental innovations.

Product and service innovations create more jobs in contrast to process innovations. Employment changes only occur in the wake of major innovations. When innovations are mainly motivated by environmental concerns, they tend not to have employment effects. Eco-innovations aiming at cost reduction reduce employment.

If the innovation is motivated by market share considerations, employment may change in both directions. We detect skill biased technological change of eco-innovations, because the share of highly qualified employees has a positive impact on employment increases, while it is insignificant for employment decreases. Additional employment is especially created in small firms. A positive correlation between sales expectations and employment found in our sample is in broad conformity with the innovation literature.

We conclude from our survey that environmental innovations have a small but positive effect on employment on the firm level. Thus, environmental support programmes do not counteract labour market policy. It should not be expected, however, that the ecological modernisation of firms can provide substantial contributions to overcome mass unemployment. A further shift from end-of-pipe technologies to cleaner production, especially towards product and service innovations, would be beneficial for the environment and create jobs. This synergy should be considered in political programmes whether they are initiated by environmental, labour market or innovation policy. Some potential still exists for shifting subsidies from end-of-pipe solutions to cleaner products and services.

The survey analyses – which use the entire international sample – are complemented by descriptions of the relevant survey results in the participating individual countries Switzerland, Germany, the Netherlands, United Kingdom and Italy.

In addition to the survey analysis, we present case studies on the effects of environmental innovations on employment in firms producing energy-saving windows or recycling refrigerators, using biomass for energy production, applying biotechnology and lending money for environmental projects. These case studies use the life-cycle assessment method in order to provide a broad picture of the employment effects that can be expected from environmentally beneficial production and service provision.

The book is based on the results of the Impact of Clean Production on Employment in Europe: An Analysis Using Surveys and Case Studies (IMPRESS) project which was financed within the Targeted Socio-Economic Research (TSER) Programme of the European Union.

The IMPRESS project was co-ordinated by the Centre for European Economic Research (ZEW), Germany. The research partners were Fondazione Eni Enrico Mattei (FEEM), Italy, Maastricht Economic Research Institute on Innovation and Technology (MERIT), the Netherlands, and Policy Research in Engineering, Science and Technology (PREST), Victoria University of Manchester, United Kingdom. Solothurn University of Applied Sciences (FHSO), Switzerland was an associated partner.

We wish to thank Sarah Gramm and Ilja Karawanov for proof reading of papers and technical editions.

The editors

Klaus Rennings and Thomas Zwick

Contents

Part I: The Synthesis Report 1

Employment impacts of cleaner production: theory, methodology and results

Matteo Bartolomeo, René Kemp, Klaus Rennings and Thomas Zwick 3

Part II: Case Studies 55

Windows

Mathias Binswanger and Daniel Schulthess 57

Recycling

Suhita Osório-Peters 83

Biomass

Andrea Marsanich 119

Biotechnology

Anthony Arundel, Ivo Demandt and René Kemp 133

Financial services

Neville Reeve and Steven Glynn 171

Part III: National Survey Results 195

The Swiss survey

Najib Harabi and Annette Jochem 197

The German survey

Klaus Rennings and Thomas Zwick 213

The Dutch survey
Anthony Arundel, Leann Chervenic-Poeth and René Kemp 229

The UK survey
Steven Glynn 267

The Italian survey
Matteo Bartolomeo and Rita Canu 283

Annex: IMPRESS Questionnaire 301

Contributors 321

Part I: The Synthesis Report

Employment impacts of cleaner production: theory, methodology and results

Matteo Bartolomeo, René Kemp, Klaus Rennings and Thomas Zwick

1 Introduction

In Europe, the protection of the environment has become an important business goal. Given the continuing high level of unemployment, the question arises whether further efforts for cleaner production may lead to a change in unemployment and if so, under which conditions. Both politicians and scientists search for strategies which help to further enhance the quality of the environment and at the same time increase employment (Bovenberg and van der Ploeg, 1994, Goodstein, 1995).

In the White Paper *Growth, Competitiveness and Employment*, the European Commission (1993: 145) identified the need "to analyse in which ways economic growth can be promoted in a sustainable way, which contributes to higher intensity of employment and lower intensity of energy and natural resources consumption". From a technological perspective, it is widely agreed that sustainable development requires a shift from end-of-pipe technologies to clean production or integrated technology (Vergragt and Jansen, 1993, Weizsäcker et al., 1995). In this context, clean production often requires both technological innovations and a "rethinking of the firm" through organisational restructuring (Gameson et al., 1997: 9).

From a micro-economic perspective, it is often argued that increasing ecological efficiency strengthens economic competitiveness and could become a key strategic factor in future international trade (Fuzzler, 1996, Porter and van der Linde, 1995). In simulations with macro-economic models, however, the existence of such a double dividend from environmental policy is still controversial (Conrad and Schmidt, 1997). The issue becomes even more ambiguous with regard to the impacts of cleaner production on employment. Some authors argue that triple dividends of greater economic growth, higher employment, and a reduced environmental burden could exist for certain innovations. For example, the trend towards increased recycling and re-use of materials leads to innovations in the field of reverse logistic systems, which then becomes a rapidly growing industry sector. In contrast, clean production reduces the demand for material, energy and end-of-pipe technologies, and thus reduces the demand for labour in these sectors. The net effect of clean production on employment in an entire economy is therefore unclear.

The ongoing scientific and political debate on the relation between environmental innovation and employment reveals that our understanding about the relationships between environmental protection, specific environmental technologies and employment is comparably little developed. In particular, empirical studies about the employment effects of integrated environmental technologies in the economy are rare. From a theoretical point of view, net employment effects can be positive as well as negative as is the case with innovations in general (see Table 1).

Environmental policy promoting integrated technologies may result in job losses in the end-of-pipe sector or lead to higher prices and therefore lower the demand for the firms' output. Reduced energy and material consumption may lead to losses in both production and jobs. On the other hand, ecological optimisation may require the simultaneous use of integrated and additive technologies (Coenen et al., 1995) limiting job losses in the end-of-pipe sector. Positive employment effects may result from improved goods quality and/or international competitiveness. Integrated technologies seem to have advantages because of their competitive edge and comparably low long-run costs for users. While necessary system changes based on integrated technologies often require higher set-up costs, less resource use and fewer investments in end-of-pipe measures save costs over a longer time horizon.

The aim of this book is to improve the theoretical understanding and empirical evidence of the relationship between integrated technologies and employment. Strategies to identify and promote triple dividends have to be very specific and selective (Gameson et al., 1997: 12). It is crucial to know which synergies and/or conflicts exist for individual countries and innovation types, and how to handle them strategically. Such specific information can only be generated by micro-studies such as surveys and case studies. In particular, the objectives of this book are as follows:

- to establish a Europe-wide methodological framework for analysing the impacts of clean production on employment,
- to use, wherever possible, data from European innovation data banks to examine the relationships between eco-innovation, employment, and competitiveness,
- to supplement existing information with in-depth case studies in selected industrial and service sectors,
- to obtain detailed, yet representative, internationally comparative information about eco-innovations through customised surveys on eco-innovation,
- to achieve a better understanding of synergies and conflicts between clean production, employment and competitiveness, and

- to provide policy makers with country-specific micro-data which can be used to help strengthen synergies and to mitigate conflicts.

Table 1: Potential employment effects of cleaner production

Potential negative effects	**Potential positive effects**
Job losses in enterprises using end-of-pipe (additive) environmental protection technologies	New jobs for brokering, consulting, and financing agencies for integrated environmental protection (cleaner production)
Job losses in the vendors of additive environmental protection technologies	New jobs for developers and suppliers of green production as well as energy and resource saving technologies
Job losses in the energy sector (production and conversion) and in resource extraction and processing due to a decrease in production resulting from an improved energy and material efficiency or lower material inputs in production processes	New jobs by "win-win options" brought about by using the saved resources elsewhere
Job losses in the investment goods industry supplying the energy and resource extracting sector	New jobs due to increased competitiveness by using production and product integrated technologies of environmental protection (cleaner production)
Job losses due to the potential increase in work productivity linked to the implementation of green production technologies	Fewer crowding out effects and higher profitability of investment in clean technologies than for the use of additive technologies

Source: Rennings and Pfeiffer (2001).

2 State of the art

The impact of innovations on employment has been analysed empirically in several studies over the past years (see e.g. Pfeiffer and Rennings, 2001, König, 1997, König et al., 1995, Smolny and Schneeweis, 1999, Rottmann and Ruschinski, 1998, Brouwer et al., 1993, Van Reenen, 1997). The studies have tested the hypotheses that:

- Product innovations have a positive impact on employment since they create new demand. Negative indirect effects occur if sales of other products decrease (substitution effect) or if prices increase due to the new product (income effect).
- Process innovations increase the productivity of firms and have a negative direct impact on employment since they are normally labour-saving investments. A negative direct effect can be compensated by positive indirect effects, if the sales of a firm increase due to lower prices.

The hypothesis that technical progress in general and process innovations in particular kill jobs has not been confirmed by the empirical data. Although the results of individual studies vary, the tendency is that innovation has a positive effect on employment and more jobs are created by product than by process innovations. Nevertheless individual studies found negative employment effects of process innovations (survey in König, 1997).

Empirical research in this area encounters some methodological problems. Firstly, when innovations are measured it is generally difficult to distinguish clearly between product and process innovations. New products often need new processes, while new processes often change characteristics of a good or service. Such interdependencies between product and process innovations are not reflected in survey data which are used in microeconomic studies. Thus the results give only a simplified picture of complex innovation processes. Secondly, surveys on the firm level measure mainly direct effects. Interpretations thus have to keep in mind that indirect effects are, if at all, only partly included.

In the political debate, increasing attention is drawn to the question of how ecological transformation towards cleaner production affects the economic performance of industries, especially concerning employment. Views about the direction of these impacts are highly controversial. A popular hypothesis is that lower inputs of natural resources in the production process due to improved eco-efficiency require higher labour inputs and thus lead to positive employment effects (Ritt, 1999: 26). This view is expressed in the slogan: "Make kilowatt-hours unemployed, not people". This position is contradicted, however, by the observation in several studies that innovations improve both energy and labour productivity and therefore replace labour. Further, it is often argued that environmental protection measures are a cost burden for domestic firms and thus weaken their competitiveness on international markets.

Within the empirical literature on employment impacts of innovation, only few studies have specifically analysed innovations which are environmentally beneficial. Pfeiffer and Rennings (2001) conducted an industry survey of environmentally oriented firms in Germany. In Austria, Kosz (1997) investigated firms being involved in environmental programmes and Köppl and Pichl (1997) analysed data from the Austrian innovation panel. All three studies found that employment impacts of environmental innovations were positive but very small. In the German survey, 84 to 91% of the firms stated that environmental innovations have no notable effect on firm-level employment (numbers differ slightly across categories of innovations). When there were employment changes, the positive ones outnumbered the negative ones. These results have been confirmed by a recent international study (Ritt et al., 2000).

3 Basic methodology

3.1 Working definitions

In this book, we use the following definition of environmental innovation or eco-innovation that has been developed in the IMPRESS project: Environmental innovations consist of *new or modified processes, techniques, practices, systems and products avoiding or reducing environmental harms*. Environmental innovations may be developed with or without the express aim of reducing environmental harm. They may be motivated by normal business goals of reducing costs and enhancing product quality. Many environmental innovations combine an environmental benefit with a benefit for the company or user. Environmental innovations are divided in technological, organisational, social, and institutional innovations. The book will, however, focus on technical and organisational eco-innovation (cf. Figure 1).

Figure 1: IMPRESS taxonomy of environmental innovation

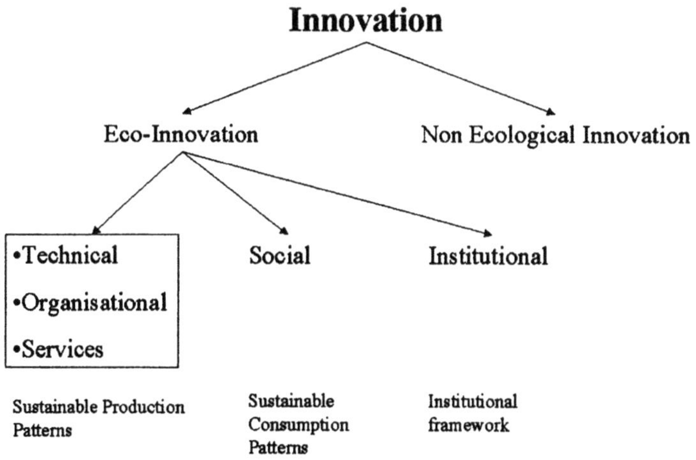

Figure 2 summarises the IMPRESS taxonomy of technical eco-innovations.

- Eco-innovations are divided into cleaner technology (product and process-integrated changes) and end-of-pipe technology (pollution control technologies that prevent the direct release of harmful substances into the environmental media air, water and soil).
- Recycling cannot easily be subsumed under the categories cleaner and end-of-pipe technology. Process-internal recycling can be understood as cleaner technology while process-external recycling is an end-of-pipe technology. To avoid any confusion, it is reasonable to treat recycling as a separate category.
- Finally, we have introduced the area of logistics, product delivery and distribution systems as a separate innovation category. Although they can be interpreted as specific kinds of process innovations, we have added them explicitly. This was motivated by the fact that the importance of product delivery, transport and distribution has increased over the past years, and not all firms may understand these activities as process innovations.

To summarise, we asked for six different categories of eco-innovations in our survey: Product integrated measures (goods and services), process integrated measures, end-of-pipe measures (pollution control), recycling, organisational measures and logistics.

Figure 2: IMPRESS taxonomy of technical eco-innovation

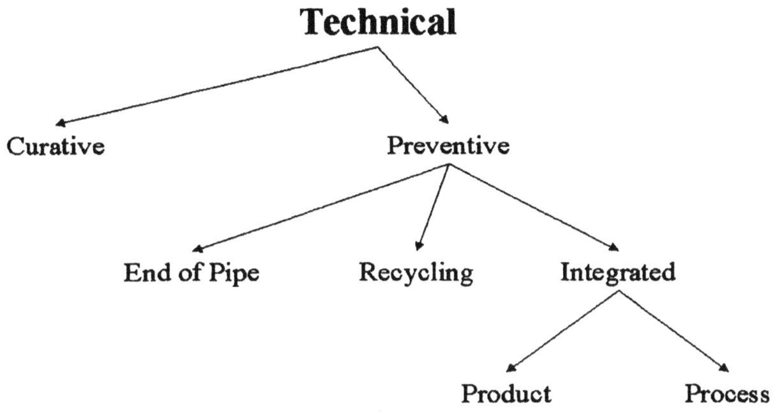

3.2 Kind of employment impacts to be studied

The employment impact considered in the case studies will be:

- direct and indirect effects,
- compensatory effects.

In IMPRESS the *direct employment effects* are defined as: effects that occur at the companies producing and using the innovation *and* effects that are directly related to the new product or process. In the case of a labour-saving new process the direct effect is negative, in the case of a labour-increasing new process the direct effect is positive. In the case of new products the direct effect is usually positive since possible substitution effects are viewed as indirect effects.

Indirect employment effects are the employment effects that occur elsewhere in the same firm (the case of a multi-product or multi-process firm where these indirect effects are related to other products and processes) or that occur in other firms. Indirect effects can be:

- *substitution effects:* reduction in employment in old processes and products after the introducing of an eco-innovation,
- *income or compensatory effects*: an increase or decrease in value added related to the production and use of an eco-innovation. They can occur in the innovating firm or elsewhere.

In IMPRESS indirect effects will be understood in a broad sense covering both substitution and income effects. As far as income effects need to be singled out as a separate category, we speak of compensatory effects.

Apart from the above employment changes, we will consider changes in skills, job content, work division, and work relations.

3.3 Hypotheses

We have formulated hypotheses concerning direct and indirect employment effects as described in Table 2. The hypotheses are in broad conformity with hypotheses in former studies on the general employment effects of innovations (see König, 1997 and König et al., 1995). We have, however, introduced two peculiarities of eco-innovation:

- Environmentally friendly products usually do not create substantially new demand. An example is the introduction of low-noise lawn-mowers. They led to more employment in the production of these devices, which are at least partly compensated by respective losses in the sale of noisy lawn-mowers. Thus it can be assumed that substitution effects of cleaner products are generally higher compared to other product innovations. Positive employment effects of cleaner products can occur if they create more value per unit. For example, the production of organic food is normally more costly than the production of conventional food which is met by a higher consumer willingness to pay for these products. Further, additional employment can be created in R&D-departments temporarily. Total employment effects of cleaner products can, however, be expected to be lower compared to other innovations.
- Environmental friendly process innovations do not necessarily increase the productivity of a firm. They may even reduce productivity and require increasing labour inputs per unit. Thus environmental process innovations can have a positive direct employment effect. These effects can be compensated by negative indirect effects, i.e. a loss of sales and competitiveness (especially if environmental standards are different across countries). This peculiarity of environmental process innovations can be explained by the fact that they are often not motivated by cost reduction and increasing sales, but by compliance with regulation (Cleff and Rennings, 1999) and can be assumed especially for end-of-pipe technologies. For example, a new filter or recycling process may be installed due to regulation. In contrast, economic innovation goals such as increased productivity and cost reduction (costs of energy, waste, disposal, material or labour) can be especially assumed for process integrated innovations and innovations in logistics.

Table 2: Hypotheses on direct and indirect employment effects of eco-innovations

Types of integrated measures	Direct employment effects	Indirect employment effects
Product integrated measures (goods and services)	Tendency positive, employment due to new product or service	Tendency negative, size depends on the degree of complementarity of old and new products and services
Process integrated measures, logistics	Increasing productivity but negative employment effect, substitution effect of technical progress	Tendency positive (compensatory effect, increasing competitiveness)
End-of-pipe processes, recycling measures	Tendency positive, due to implementation and operation of new technology. Increasing productivity is not the main innovation goal (other motivations such as compliance with regulation dominate)	Tendency negative (compensatory effect, loss of competitiveness)
Organisational measures (e.g. eco-audits)	Positive employment effect due to implementation of the organisational measure	Unknown, depends on concrete measures within environmental programme of the firm

Product integrated innovations, eco-friendly services

Product innovations in integrated environmental protection lead to positive direct employment effects, which can, however, be partly or entirely offset by their crowding out of previous products. Our study is the first one looking specifically at the effects of environmental innovations in the service sector, too. Our working hypothesis is that the effects are similar to those of product integrated measures in other economic sectors.

Process integrated measures, logistics

Our hypothesis is that the employment effects of process integrated environmental measures are in terms of their employment effects comparable to the cost-saving technological progress by other process innovations. In addition to the ecological effects, increasing productivity is sometimes the main reason for the innovation. As competitiveness improves, the indirect effects inside the firm tend to be positive. Negative indirect effects outside the firm may occur in other sectors due to reductions of waste, transport, energy and material use (job losses for supplying firms in the waste, energy, transport and production sector). The same is expected for environmental logistics innovations which often include measures to reduce transport.

End-of-pipe processes, recycling

End-of-pipe and recycling measures tend to have positive direct employment effects. They create new steps and links in the value chain and thus have a potential for additional employment. They are normally accompanied by additional investments which do not necessarily increase the productivity of the firm. The indirect employment effects tend to have the opposite sign. The employment effects are thus the opposite of the expected effects of integrated process innovations.

Organisational measures

Organisational measures are initially accompanied by additional expenditure and work processes (e.g. undergoing an eco-audit procedure), which create positive direct employment effects, while the indirect effects depend on the concrete measures which are implemented within the environmental programme of a firm.

It should be noted that our firm survey mainly measures direct effects within the firm. Thus empirical evidence drawn from our study focuses on our corresponding hypotheses on direct effects. However, some questions have been included in the questionnaire which allow interpretations concerning the relevance of indirect effects. For example, we asked for the effects of environmental innovations on prices. Substantial price effects can be used as an indicator that indirect income effects may be significant.

4 Case study methodology

Case studies in the IMPRESS project had the aim to distillate a methodology for understanding the relationship between employment and cleaner production in specific situations. This understanding is necessary as a basis for certain policy decisions, e.g. if cleaner production should be supported by public programmes at all when employment effects are considered, or if certain technologies are preferable from an employment perspective.

The methodology is based on the hypothesis that most of the employment effects related to a specific innovation occur along the value chain of the adopting firm. The methodology can be used by policy makers interested in promoting employment and environmentally friendly policies. The methodology is also interesting for researchers operating in the domain of innovation, employment and environment.

A second objective of the case study exercise was to validate IMPRESS project hypotheses and to provide a first understanding of the different variables influencing the relationship between employment and cleaner production and the results therefore serve as input for the design of the survey questionnaire in the IMPRESS research.

It was not the aim of the case studies to provide comparable data on the relationship between employment and cleaner production. It is important to keep in mind that case studies have been carried out in conjunction with a large survey involving around 1,600 companies in five European countries.

4.1 Application of the life-cycle assessment

The methodology behind the case studies is inspired by the life-cycle assessment[1] (LCA) methodology. This is primarily used by companies wishing to compare different alternatives for optimising environmental products performance (in order to communicate environmental attributes of products) or to compare product characteristics to certain standards such as those defined in eco-label schemes. Companies can also use LCA, or simplified versions of it, in order to improve the environmental profile of purchasing activities.

LCA is also used by local and national governments as a decision support tool to compare different policy scenarios in the domain of environmental protection and industrial policy.

The basic concepts behind life cycle assessment are that the environmental performance of a product is always relative (to another product with a similar function) and it is always the result of a number of environmental impacts that occur from the initial stages of acquisition of raw materials to the final disposal of the product, including all the intermediary steps of processing, transporting, distributing and consuming the product itself and all the materials needed.

Similarly to the LCA, the methodology used in the IMPRESS project consists of four key steps:

1. **Initiation**. This includes the identification of:

- the characteristics of the innovation to be studied,
- system boundaries for the assessment of employment impact (especially relevant for the value chain and local/regional context analysis),
- kinds of employment effects to be analysed (direct, indirect, micro, meso, macro).

2. **Inventory**. This includes the assessment of employment effects on all the phases and actors chosen according to the definition of system boundaries. The assessment should at least identify the kind of impact (positive or negative) for

[1] A concise and easily accessible document on LCA has been published by the European Environmental Agency (1997).

each element considered within the system boundaries and possibly provide a more precise and quantitative evaluation of the employment effect.

3. **Impact assessment** and
4. **Interpretation** of results. This includes a classification and categorisation of employment impacts and a discussion of the findings according to key variables such as time frame, geographical scope, possible technology breakthroughs, changes in demand and clients/consumer perception.

4.2 Initiation

The object

The first step is the identification of **eco-innovations** (technical, organisational, service) within the chain to be examined. The innovation should be **implemented** and not just designed. In many examples it will be difficult to separate, for example, the technical innovation from other motivations. Frequently a technical innovation is accompanied by an organisational innovation. Both will be studied, but one will be the **main object**, while the accompanying innovations will represent an **effect of the main innovation**.

Example solvent-free paint:

- eco-innovation: solvent-free painting process in the car industry
- related innovation (organisational): training and organisational measures to manage the innovative process
- related innovation (product/service): car with solvent-free paint.

The functional equivalence

In the LCA methodology, environmental effects of a certain product or service are compared with a functionally equivalent unit. This means for example that a product is compared with another one able to deliver a similar function. The choice of the appropriate reference is crucial in judging the value of the innovation.

Similarly, for our purposes, we also need to identify a functional equivalent for the innovation that will be studied in order to understand the employment effect.

1. The **simplest way**, but not necessarily the most meaningful one, is to compare the existing situation with the situation before the innovation was introduced. By doing so, employment effects will be studied on new and traditional factors involved in the chain.
2. The other possibility could be to identify a similar context where this innovation **did not take place**.

3. A third possibility, probably less feasible, is the identification of a **theoretical reference**.

Option 1 seems to be the most cost-effective approach.

The system boundaries and the employment effects

System boundaries should be clearly defined before starting the inventory phase. Nevertheless, system boundaries may also be amended during the case study in order to cope with unexpected issues and to provide useful methodological feed-back.

The identification of system boundaries will take into account various criteria, such as:

- the kind of eco-innovation to be studied
- the integration of the production chain
- the possibility of collecting useful and correct information.

There will always be trade-offs between the quality of information and the scope of the boundaries.

System boundaries should also be described in a **flow chart**. The elements of this flow chart will be the elements playing a role in development and adoption of the eco-innovation and also the elements at the end of the life cycle of the product manufactured by the company adopting the eco-innovation.

Example - Some elements to be included in system boundaries:

- car manufacturer (various employment effects within the same organisation)
- developers of painting processes (new and traditional)
- paint producers (new and traditional)
- waste management for paints (new and traditional)
- car maintenance service (for car bodies)
- waste management for end of life cars (new and traditional)
- consultants and auditors.

4.3 Inventory

For each of the elements considered within the system boundaries, the researcher should assess:

- the variations in the number of days worked and

- the skill effect.

This assessment applies to elements involved before and after the introduction of the eco-innovation. When making such an assessment, it should be clarified if the effects refer to the specific period of time when the innovation is adopted or if they are of a more stable nature. The overall inventory will therefore have two different dimensions:

- a detailed assessment of qualitative and quantitative employment effects in the short run, at micro (and meso) level. This will be also presented in a graphical way using a flow chart similar to that related to the initiation phase.
- an aggregated evaluation of the employment effect in the short run, at micro (and meso) level.

4.4 Impact assessment and interpretation

Here the more qualitative findings of the case study phase will be discussed.

In particular, the impact assessment and interpretation will make use of the findings presented in the inventory phase:

- reliability of information collected
- strengths and weaknesses related to system boundaries
- possible effects in the long term (beyond the short term)
- key determinants for positive and negative effects
- generalisation of results to similar chains
- discussion of possible macro effects
- discussion of geographical distribution of effects
- possible policy implications
- comment of the methodology.

Moreover, this phase will explicitly refer to general research and case study specific hypotheses and will comment them.

5 Survey methodology

IMPRESS used a common concept for a stratified sample in all countries. The following agreements were made:

- only firms with more than 50 employees are considered,

- some sectors are excluded (health and social work, public administration),
- a stratified sample is drawn containing 8 cells (small firms with 50-200 employees, big firms with more than 200 employees; industry firms, service firms, other cells are defined individually by each partner, e.g. regional differences such as East-West in Germany and South-East in Italy),
- 400 interviews with eco-innovative firms were realised in big countries (Germany, UK, Italy), 200 in small countries (Netherlands, Switzerland).

Between July 1999 and February 2000, 10 different versions of the questionnaire were drafted and discussed. Before the questionnaire was used in the field phase, the development phase included 2 pretests (January and March 2000) and a methodology workshop with international experts (February 2000). The final questionnaire is enclosed in the Annex.

All countries used the same methodology, the same concept for the stratified sample and the same questions. The large amount of time and discussion invested in the formulation of the questionnaire in order to ensure applicability in all countries was justified because no major problems occured during the pretests and the field phase. The descriptive and econometric analysis showed that the data can be easily compared between countries. IMPRESS therefore has created a unique European databank on environmental innovation and employment.

Between March and July 2000, 1,594 telephone interviews with industry and service firms were carried out in five European countries (401 in Germany, 384 in Italy, 201 in Switzerland, 400 in the United Kingdom, 208 in the Netherlands). The addresses for the telephone interviews were drawn from a stratified sample with the dimensions small firms (between 50 and 199 employees) and large firms (200 or more employees) and 8 sectors according to the NACE codes D-K. These NACE codes comprise industry, manufacturing and services. Firms active in other sectors such as mining, agriculture or public administration have not been included in the sample.

In Germany, an additional stratification for the firms located in East or West Germany was introduced, in Italy, the firms were differentiated between the North and the rest of the country, while in Switzerland, a differentiation between the regions of the three major language groups German, French and Italian was made.

The firms contacted were asked first if they introduced at least one eco-innovation during the last three years. If this was not the case, the interview was terminated. Therefore, the data basis only contains firms that identified themselves as eco-innovators and the analysis concentrates on the behaviour of firms with respect to employment changes provided they have introduced an environmentally related innovation. In addition, we only have cross-section data and therefore neither an analysis of causal effects nor of the impact of time is possible. Individual fixed effects cannot be taken into account. The time structure is only captured indirectly by asking about the employment impact during the last three years and by asking about the expected demand effects (see also König et al., 1995, Rottmann and Ruschinski, 1998). These limitations of the data seem to be minor, however, because we do not estimate employment change equations or labour demand, but only impacts of different eco-innovations on the sign of the employment change. A thorough discussion of possible data limitations and related estimation problems can be found in Chennels and Van Reenen (1999).

The number of small and large firms and the number of firms interviewed per sector are reported in Table 3.

Table 3: Description of the sample

	Number of firms	Share in %
Small	1,203	75.47
Large	391	24.53
Industry/manufacturing (NACE-Codes D-F)	906	56.84
Of which: manufacturing	736	46.17
Electricity, gas and water	33	2.07
Construction	137	8.59
Service (NACE-Codes G-K)	688	43.16
Of which: Wholesale/retail-trade	263	16.50
Hotels and restaurants	37	2.32
Transport, storage and communication	156	9.79
Financial intermediation	61	3.83
Real estate, renting and business activity	171	10.73

We used a stratified representative sample considering the cells mentioned above. The results of the survey are therefore representative for each country under the assumption that eco-innovators do not differ in their characteristics from other firms. Since this is a very restrictive assumption, the survey results are not representative for all eco-innovators. A representative survey of eco-innovators can only be carried out if the universe of eco-innovating firms is known, which is not the case. Determining the universe of eco-innovating firms and improving the knowledge about general differences between eco-innovating and other firms is beyond the scope of our project and remains as a question for further research.

The data set was especially designed to measure the effects of eco-innovations on employment at the firm level. Therefore, it contains some unique variables that are not included in other data sets. For example, it directly asks about the employment effects induced by the innovation in contrast to the general employment change which is frequently used as an indirect indicator for it, see for example Pfeiffer (1999). In addition, besides the differentiation between direct and indirect employment effects, the data set allows to draw conclusions on the employment effects of relevant policy variables such as subsidies and environmental regulations.

6 Results

6.1 Case studies

The case study analysis in the IMPRESS project focused on selected value chains where a certain environmental innovation has taken place. In the case studies the adopter of the innovation is viewed as a central element and as a starting point from which to consider upstream and downstream effects related to the specific innovation being examined.

The selection process for the eco-innovation considered a wide range of relevant innovations as described in the general approach, e.g. process and product innovations as well as innovations in the manufacturing and in the service sector. A specific case study was carried out for recycling since recycling was characterised as a specific innovation category that does not fully fit either in the category integrated technologies or in the category end-of-pipe technologies.

Using the taxonomy of eco-innovations adopted by the IMPRESS project, environmentally beneficial innovations studied in this phase can be grouped as follows:

- **Product integrated environmental innovations**, such as the production of double glazed windows studied by the University of Applied Sciences, Soloturn (FHSO) in Switzerland,
- **Environmental service innovations,** such as green loans in the financial services industry – studied by PREST (University of Manchester) in the United Kingdom,
- **Process integrated innovations**, like the generation of energy from biomass, as studied by Fondazione Eni Enrico Mattei, Milan in Italy, or like the use of enzymes in industrial biotechnology application and the use of genetically modified crops in agriculture (both are part of biotechnology applications) studied by MERIT, Maastricht in the Netherlands,
- **Recycling innovations**, like the two-stage recycling of refrigerators as studied by ZEW, Mannheim in Germany.

Organisational measures and end-of-pipe innovations have not been taken into account in case studies. A case study on organisational innovations has already been carried out in a previous project (see Pfeiffer and Rennings, 2001). End-of-pipe technologies were not considered for case studies since the focus of the project is cleaner production.

The case studies are briefly described below in order to provide the reader with basic elements for understanding the following discussion on employment effects and methodology for studying it.

6.1.1 Production of energy-saving windows

The case study analyses the value chain associated with the production of windows with clear environmentally friendly features. The specific innovation is placed in the broad context of construction business, which is one of the most labour-intensive sectors of western economies and always considered as key target for governments' employment policies.

At the same time the construction sector is well-known for its negative environmental impact related to the nature of the business and to the size of firms, making the enforcement of environmental legislation frequently ineffective.

Energy-saving windows, while not solving the problem of the eco-profile of the construction process, are seen as outputs of the construction activity that have a positive environmental impact on the final product of the process: a new or restored building.

The value chain of modern windows is much more complex now compared to only few years ago, when windows were simply made of wood, glass, some metal parts or similar material. Today the consumer can choose among various options with a combination of materials like: wood, aluminium, PVC, reinforced glass, quartz and others. In addition, different sorts of gasses are used in double or triple glazing windows in order to improve thermal insulation. The downstream phases are also quite articulated involving normally a number of contractors that take care of transport, distribution and installation on site and associated services.

The value chain approach here indicates that the employment impact, in the case of production of innovative windows, occurs at the level of the adopting company and at the downstream level, where the activity for transportation and installation of windows is very labour-intensive and also in relation to the customisation of the window size to consumer needs.

The assessment of employment effects should be related to the overall development of the construction business, which is by nature very cyclical and therefore becomes a key determinant for the possible fluctuations in the demand of windows. The case study suggests that the increase in employment for the company adopting the innovation is even more impressive if the sector fluctuations are taken into account.

The reference situation (the functional equivalence) is difficult to identify when the innovation is widespread as is the case of energy-efficient windows that now dominate the market. Environmental regulation was identified as the main driving force for the diffusion of energy-saving windows in Switzerland. The use of this environmentally friendly product is obligatory in new buildings in Switzerland due to environmental legislation.

6.1.2 Recycling of refrigerators

The case study considers an innovation introduced by a German firm in order to recycle obsolete refrigerators. In Germany recycling of electrical and electronic scrap (EES) is an important regulatory priority and a business opportunity. Given the size of the problem, the environmental concern, the organisational and the technical difficulties, the recycling of refrigerators is an interesting case and an instructive example for the recycling industry as a whole.

The innovation considered, a two-stage process that allows both the recovery of material and ChloroFluoroCarbon (CFC), is environmentally friendly in absolute and relative terms and it compares well with various alternatives that do not offer the same environmental gains. In particular, the innovation is based on a mechanical process that substitutes for, amongst others, the more labour-intensive manual dismantling process.

Thus, the innovation is not the introduction of a recycling system where it was missing, but an improvement in recycling practices.

The case study considers the value chain in order to assess the direct and indirect employment effect, which is then calculated taking into account two different reference situations: the first compares the situation before and after the introduction of this specific innovation. The second considers a similar context where this innovation did not take place.

The study shows that most of the positive employment effects occur in the upstream phase, collection and dismantling of refrigerators, which is also the most labour-intensive one. The direct employment effects, occurring at the level of the recycling company, are negligible, especially if we exclude the R&D and plant construction activities, that are one-off. Employment effects are also negligible in the downstream phase, which is mainly related with the re-use of materials.

As in the case study of energy-saving windows, environmental regulation was identified as the main determinant of employment in recycling firms. Labour demand largely depends on the recycling standards being defined in the regulatory process. The case study shows that in the long run labour-saving technologies are also developed in the environmental sector, i.e. positive employment impacts of new environment technologies decrease over time due to rationalisation processes.

6.1.3 Production of energy using biomass

The environmental impacts related to the production, distribution and consumption of energy are well-known and significant in all economies. Research on alternatives to non-renewable resources is going in various directions from wind to solar, from small scale hydro to biomass.

The production of energy through biomass includes a wide range of options, all of them based on the use of biomass as direct fuels for processes or as raw materials for the production of fuels by using partial combustion, fermentation, bacterial digestion or gasification. The use of biomass can then lead to heat, electricity or fuels (charcoal, alcohol, biogas, natural gas substitute).

The environmental benefits are clear: less use of non-renewable resources, use of biomass that otherwise would be wasted, and use of local sources for heat or power generation. These benefits, linked with economically viable technologies, are now attracting a lot of attention from policy makers, local authorities and entrepreneurs. While the installed capacity is very limited, the potential appears to be promising for making these processes a partial alternative to traditional energy generation.

The employment impact, according to different studies, can be positive. This case study is conducted to assess, using a value chain approach, the possible employment implications of a heat generating plant (and a heat distribution network) located in the north of Italy that uses sawmill residues as fuel. This fuel is a by product of the wood industry and, to a lesser extent, biomass produced by local agriculture.

The plant, which has a fuel oil unit that allows it to cope with peak demands, has been recently constructed using the best available technology. It allows for a continuous control of customers' requests in order to regulate the amount of energy to be produced and delivered throughout the day, limiting the overproduction of energy.

In order to assess the employment impact, the current situation with the biomass plant has been compared with a past situation characterised by a single heating system installed at the household level.

The net employment effect of the innovation along the value chain (which considers upstream and downstream activities) can be considered positive if a time frame of 10 years is the timeline for the calculation and if the construction phase for the plant and the pipeline for the distribution of heat is taken into account. However, the positive effect may be offset by the negative employment effect due to reduced maintenance of household heating plants and distribution of diesel.

In addition to the positive effects described above, the case study suggests that other positive employment effects can occur at the local level due to positive income effects on the local economy.

6.1.4 Applications of biotechnology

After providing a useful introduction to biotechnology, the analysis takes into account four different types of environmentally beneficial innovations using biotechnology. The choice of four mini-cases instead of one in-depth case was made for two reasons. On the one side, the relatively recent introduction of biotechnology in some sectors implies difficult access to consistent information. On the other side, the public concern related to some applications of biotechnology, namely genetically modified organisms (GMO), has had an impact on the willingness of companies to co-operate with researchers.

The mini-cases can be divided into two different categories. The first is represented by the use of enzymes in industrial processes with three cases addressing different industry sectors: pulp and paper, starch conversion and modification, fine chemicals. The second considers the use of biotechnology in the agro-food industry and in particular the use of genetically modified seeds in agriculture.

In the first set of cases, those focusing on enzymes, the environmental benefits are very relevant and clear and include: substantial improvements in energy efficiency; reduction in the use of chemicals and other inputs; and a reduction in waste production.

In the example of modified seeds, the environmental gains can also be very relevant (less use of herbicides, pesticides, and water), but public concern regarding possible environmental risks suggest a more cautious judgement on these aspects.

The employment effect is analysed using direct interviews and, at the same time, data available from recent surveys. The novelty character of many of these innovations prevents the employment effect for a mature market to be precisely determined.

In all cases considered, the employment effect is calculated at the level of innovation adopters and related downstream and upstream activities. The overall employment effect is reported to be negligible or negative as a consequence of increased reduction of inputs (material, water, energy) and reduction in waste streams. Another reason for the poor positive employment effect can be attributed to the high degree of automation of enzyme production thus meaning that the switch from chemicals to enzymes does not produce positive employment effects at the upstream level.

A positive employment effect can occur in situations where R&D activities are also carried out at the level of innovation adopting firms.

6.1.5 Lending money for environmental projects

Unlike the previous case studies, which focused on industry sectors, this case study considers recent innovations in the services industry, namely the financial sector. The financial sector is increasingly engaged in reducing its direct and indirect environmental impact through:

- process innovations, for example by optimising the use of energy (for building heating, cooling and lighting) or by downsizing paper consumption
- service delivery innovations, by adopting alternative ways to deliver financial services for example through telephone banking or internet banking
- product innovations, for example by offering new financial products with environmental features.

The case study in particular addresses the last typology of innovation (product innovation) and analyses two different banks in Great Britain that are engaged in lending money for environmentally friendly projects.

As far as the employment effect is concerned, the case study analysis has underlined that most of the employment effects occur at the level of the beneficiary of the loan, while the effects for the financial firm are negligible. These effects are positive but further consideration needs to be given to whether they would have occurred even without the innovation since, according to the financial firms interviewed, the service offered is viable from an economic point of view and therefore the specific innovation captures a market that could be addressed by competitors offering traditional products.

From a methodological point of view the case study suggests that the value chain approach can be usefully applied, but a number of practical difficulties may arise. First of all, the employment effect in theory should refer to: the direct employment impact at the financial firm level, the indirect employment effect at the level of competitors of the innovative financial firm, the indirect employment impact at the level of beneficiaries of the loan, and the indirect employment impact at the level of suppliers and customers of the beneficiary of the loan.

6.2 Survey results

6.2.1 Environmental innovations and employment

Figure 3 shows all environmentally beneficial innovations according to our definition which have been introduced by the firms in the last 3 years (column 'Mentioned'; multiple answers were possible). The figure also shows the innovation which has been cited as the most environmentally beneficial one ('Most beneficial', here also multiple categories were given by some firms). The most environmentally beneficial innovation is the one the entire questionnaire is referred to. Therefore, if the firm has introduced more than one innovation, the respondent had to choose the one that had the highest positive impact on the environment for the interview. Besides the innovation types "process and product integrated environmental innovations", also recycling and pollution control (end-of-pipe technologies) have been frequently introduced. Changes in the distribution system (logistics) and in organisation methods are not widespread.

Figure 3: Environmentally beneficial innovation

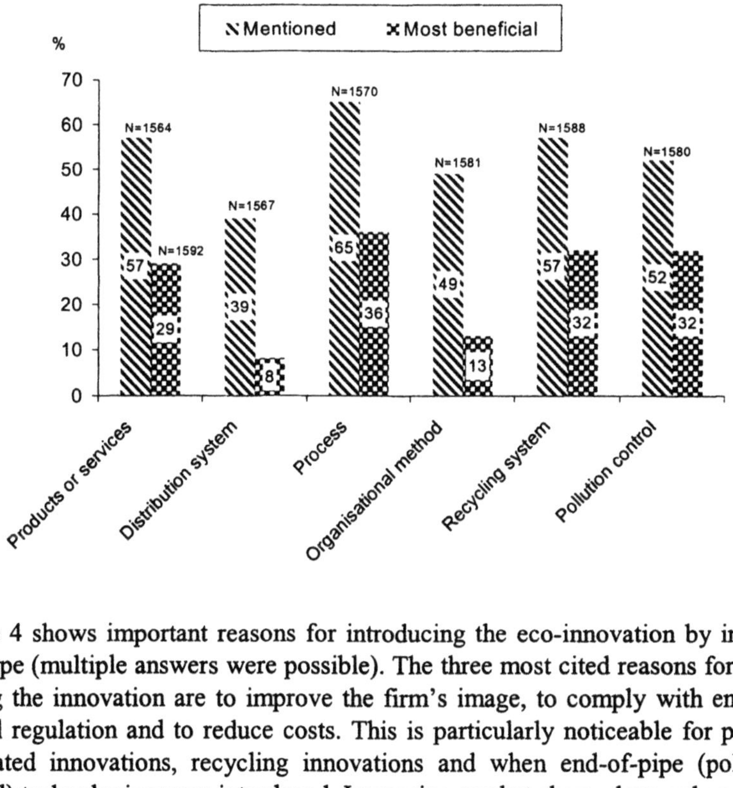

Figure 4 shows important reasons for introducing the eco-innovation by innovation type (multiple answers were possible). The three most cited reasons for introducing the innovation are to improve the firm's image, to comply with environmental regulation and to reduce costs. This is particularly noticeable for process integrated innovations, recycling innovations and when end-of-pipe (pollution control) technologies were introduced. Increasing market share plays only a minor role for introducing eco-innovations but is particularly important for integrated technologies (product, service and process integrated).

Figure 4: Important reasons for introducing the innovation

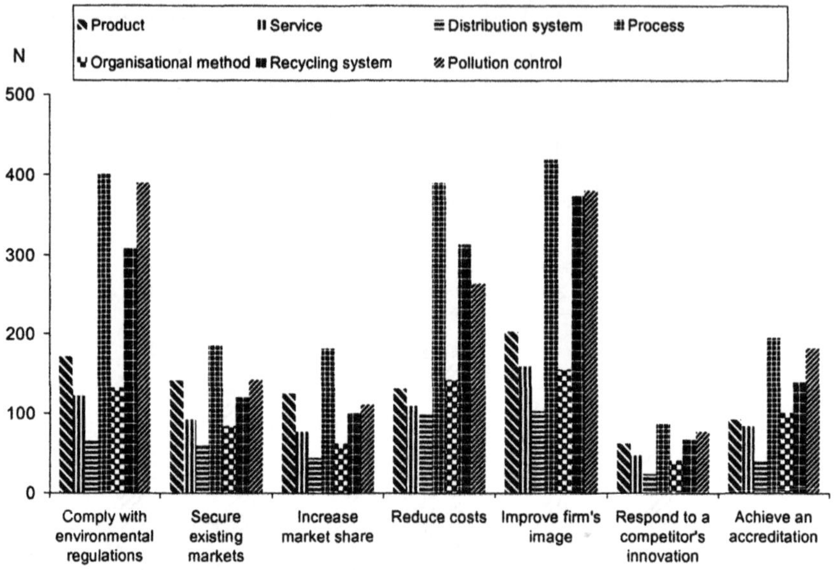

Figure 5 shows the effects of the environmental innovation on employment. Overall 88% of the firms report no notable effect on employment due to the specific innovation (see left column). In 9% of the cases the number of long-term employees increased due to the innovation, in 3% of the cases it decreased. This shows that there is only a weak but positive relation between the introduction of environmental innovations and employment.

Regarding the distribution of employment effects by innovation type, it becomes apparent that product innovations and service innovations have a sizeable above-average positive employment effect (18% and 20%). It is further interesting that the employment effect of recycling innovations is positive in almost all cases. Innovations in logistics have the highest shares of negative employment changes.

Figure 5: Effect of eco-innovation on employment

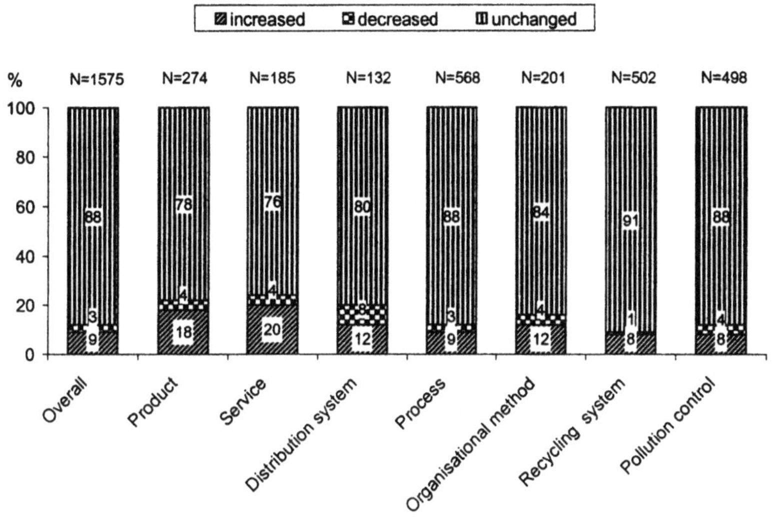

6.2.2 Factors influencing employment effects of eco-innovations

Figure 6 shows by whom the eco-innovation was developed, overall and in connection with the question whether the innovation resulted in employment effects. For 44% of the firms, other firms or institutes developed the innovation (see left column), while only 32% developed the innovation in their own institution. 39% to 40% of the firms with employment changes due to eco-innovations, developed the innovation themselves. This means that in-house development of eco-innovations results more frequently in employment effects.

Figure 7 gives an overview of the investment costs and their correlation with the employment effect of the eco-innovation. 64% of the firms invested less than € 50,000. The 135 firms which increased employment have an above-average level of investment costs. Firms which decreased employment due to the innovation also have an above-average share of investment costs above € 50,000, i.e. it seems that innovation size has an impact on the variance of employment effects. This is plausible if we take into account fixed employment turnover costs. Minor changes in the production process are therefore not employment effective.

Figure 6: Who developed the innovation

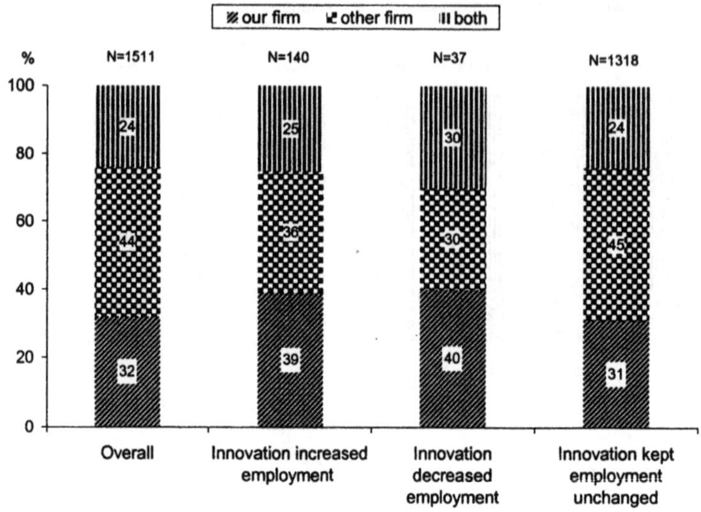

Figure 7: Investment costs of the innovation

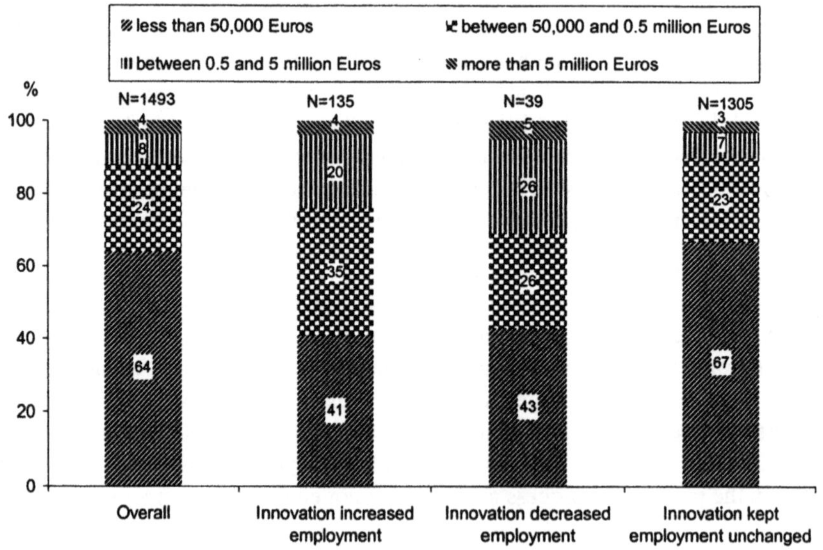

Figure 8 shows which proportions of the establishment's total innovation expenditures over the past three years were spent on environmental innovations. For the majority of establishments (51%), less than 5% were spent on the eco-innovation. Splitting the results by innovation types, it is remarkable that an above-average number of firms had a relatively higher investment share for product, service and process integrated innovations. When comparing this result with Figure 3, there is no clear connection with the employment effects of such innovation types, however. Compared to process integrated innovations, product and service innovations have a small, above-average positive effect on employment. After the observation that more expensive innovations more frequently induce employment changes, it is not surprising that innovations with a high share in total innovation costs also induce employment changes more often. While only 14% of the firms where employment was unchanged by the eco-innovation reported an eco-innovation share above 50% of total innovation expenditures, it was 42% of the firms reporting employment increases and 16% of the firms reporting employment decreases.

Figure 8: Proportion of innovation expenditures

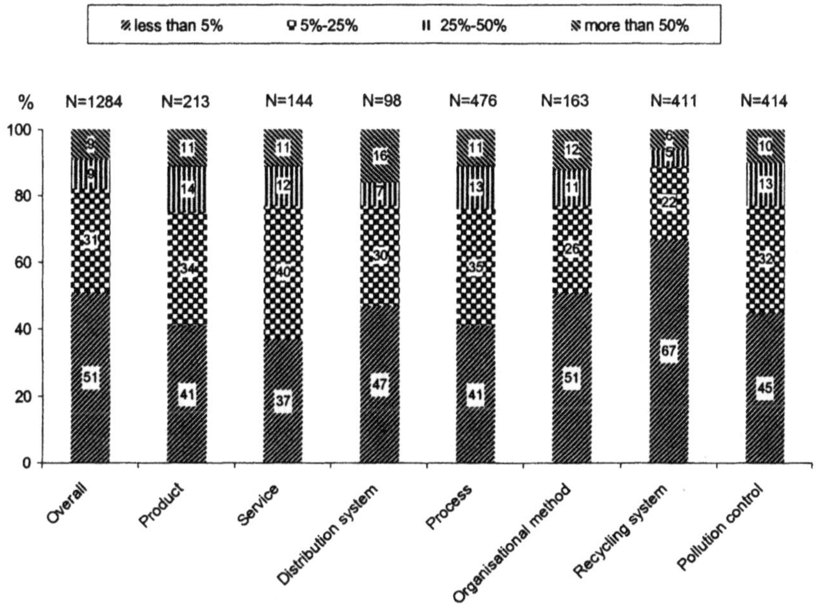

Figure 9 gives an answer to the question whether there is a correlation between receiving subsidies or grants for introducing the innovation and the innovation type. Multiple answers were possible. On average, only 11% of the firms received a subsidy or grant. For product and service innovations, the number of firms is above average. Since exactly these innovations are the ones with the strongest positive employment effect, the overall allocation of subsidies within environmental technology support programmes in the five countries involved can be characterised as employment-friendly. This also can be seen from the fact that 21% of the firms indicating employment increases due to the innovations received subsidies, while this was the case for only 7% of the firms that reduced employment. The share was 9% for the firms that did not change employment.

Figure 9: Was subsidy/grant received (by innovation type)

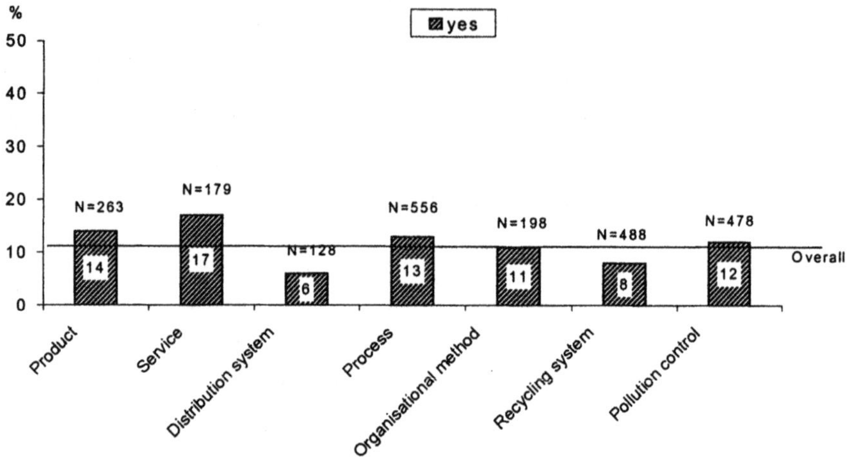

6.2.3 Eco-innovations and skills

Figure 10 shows the answer to the question whether the new skills required because of the eco-innovation are correlated with the innovation type and with employment. Overall, new skills were required by about 36% of the firms introducing environmental innovations. It can be shown that service innovations (50%) and organisational methods (46%) as well as product (45%) and process integrated (44%) innovations require a noticeable above-average need for new skills. Furthermore, there is a strong connection between the need for new skills and a positive employment effect: 67% of firms with newly hired long-term employees due to the innovation reported had a need for new skills. On the other hand, there was also an above-average need for new skills in firms which reduced employment due to the innovation.

Figure 10: Need for new skills by innovation type and employment effect

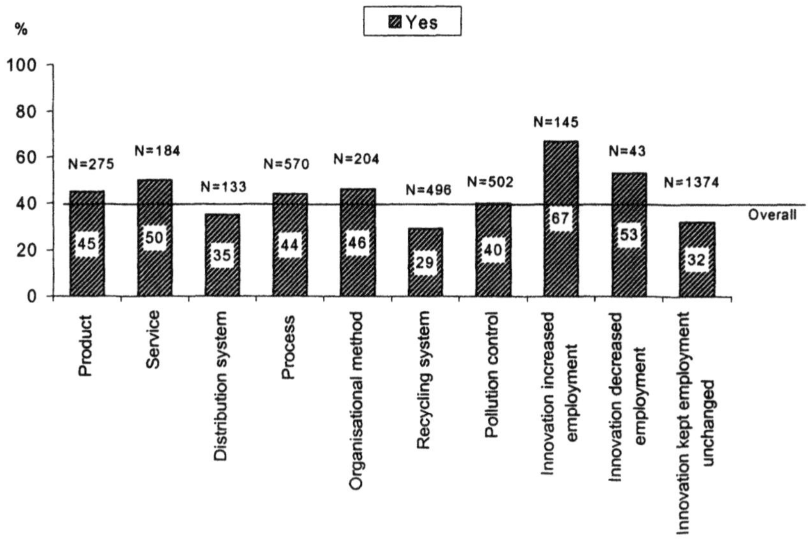

Figure 11 gives an overview how the need for new skills was met (multiple answers were possible). The majority of the firms (90% of firms requiring new skills; 41% of all firms) uses the method of training existing employees. Only 14% of all firms (33% of firms requiring new skills) hire new employees on a permanent or temporary basis. From the 10% (17% of firms requiring new skills) which outsourced this work, it can be expected that positive employment effects occurred in the firms commissioned.

Figure 12 gives an answer to the question who received training and whether this depends on sector affiliation or the employment effect. Overall, mostly skilled workers received training (45%), followed by staff with college or university degrees (32%) and by unskilled workers (24%). This distribution is similar for both manufacturing and service firms. In firms which increased employment, employees received above-average training, especially the staff with college or university degrees. In firms which decreased employment, unskilled workers and staff with college or university degrees received the same share of training (29%).

Figure 11: How were new skills met

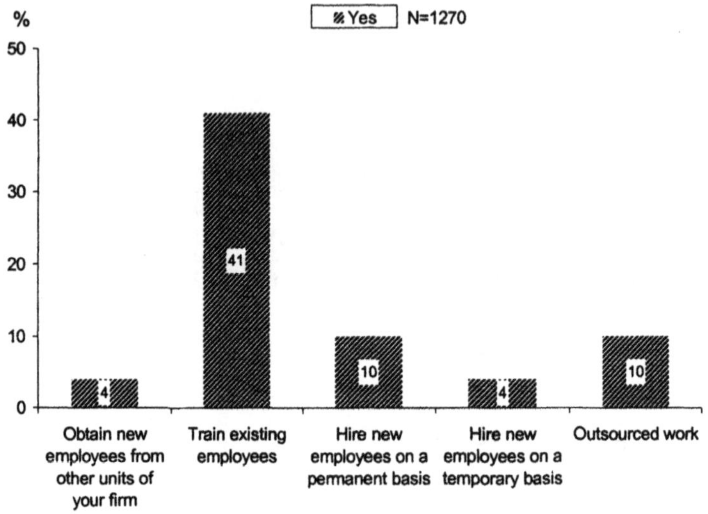

Figure 12: Who received training, by industry and employment effect

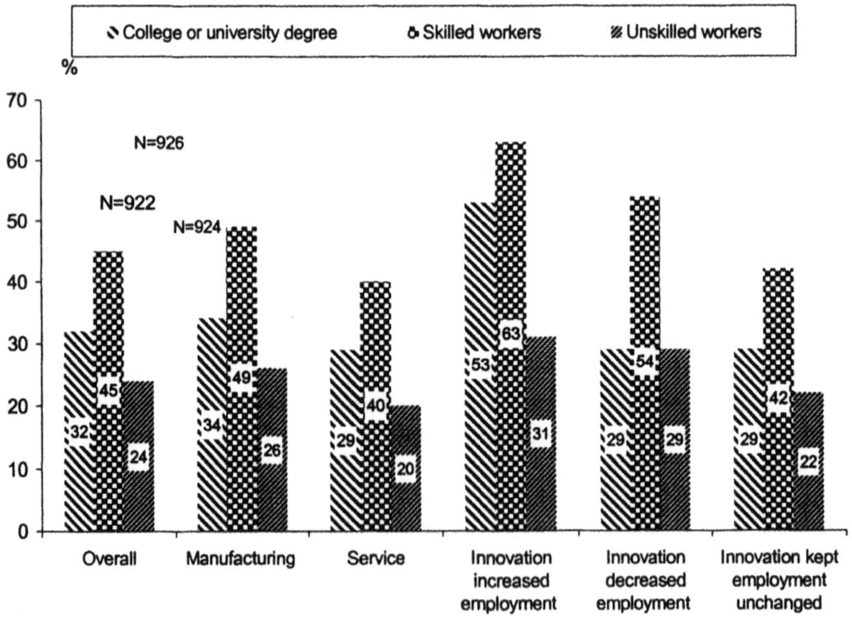

6.2.4 Effects on sales, prices and costs

The following figures show the effects of the environmental innovations on sales, prices and costs. They therefore allow us to make tendency statements about the indirect innovation effects. As can be seen in Figure 13, the innovations had no effect on sales and prices for 82% of the establishments. For 16% of the establishments, sales increased. Prices increased in 9% of the firms but in most cases by less than 5%. Prices decreased also in 9% of the firms. Since 82% of the firms state no price effect and for the rest neither negative nor positive effects dominate, it can be concluded that indirect income effects stemming from the innovation can be neglected on the firm level.

Figure 13: Effects of innovations on sales and prices

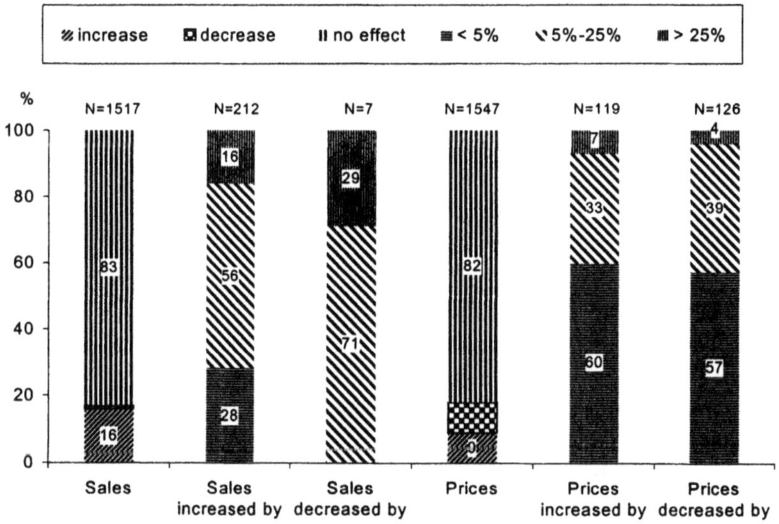

Figure 14 shows that energy costs decreased in more firms than the other costs listed here. 34% of the firms could decrease their energy costs due to the innovation. This result is in line with the high share of process innovations (see Figure 1) and the importance of the motive of cost reduction (see Figure 2). The level of decrease of energy costs is also relatively high: for 52% of the firms, energy costs decreased by more than 5%. Material costs were unaffected for 58% of the firms. In those cases with changes in material costs, the number and level of decrease were a bit higher than in the cases of increase. While substantial decreases of energy costs indicate negative indirect employment effects for the energy supplying industry, no similar effect can be found for materials.

Figure 14: Effects of innovations on energy and material costs

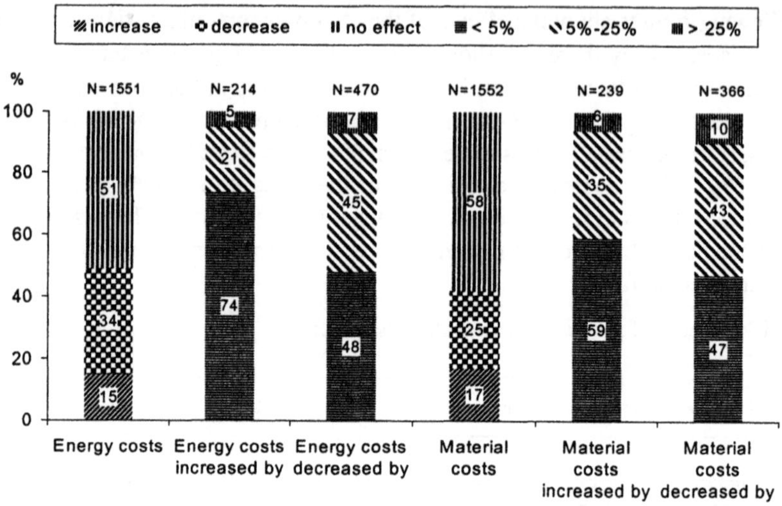

Waste disposal costs could be decreased for 37% of the firms (see Figure 15). This is due to the relatively high share of recycling innovations. The level of decrease was even higher than 25% for 22% of the establishments. However, waste disposal costs increased for 19% of the firms. Thus a total negative indirect effect can be observed for employment in the disposal sector. In 65% of the firms, labour costs did not change due to the innovation. However, they increased for 20% of the firms but mostly by less than 5%. Compared to the increased or decreased employment (see Figure 5), the higher percentage of firms with an increase in labour costs can be explained by the fact that labour costs also increased due to the need for new skills and employees with higher qualifications.

Figure 15: Effects of innovations on waste disposal and labour costs

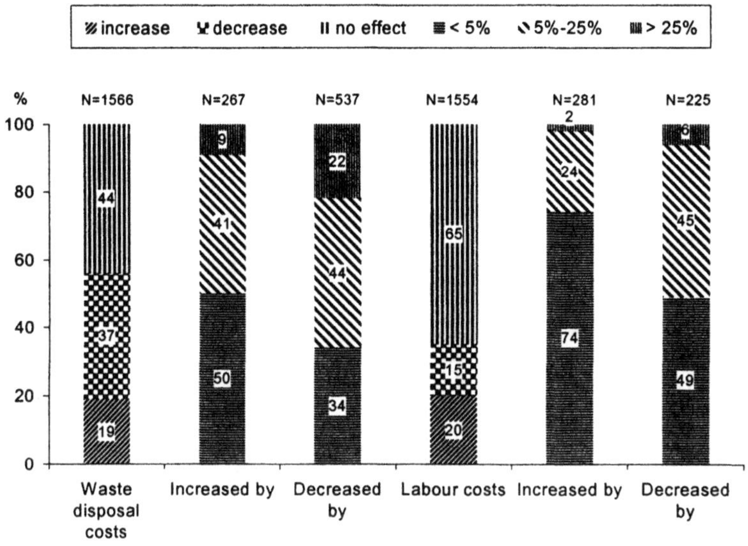

6.2.5 Other factors

Figure 16 gives an answer to the question how important are environmental regulations for processes and products and whether concrete changes were made in the last three years in order to comply with these regulations. It becomes apparent that environmental regulations seem to be important for both product and process innovations. Concrete changes in order to comply with the regulations were made in about half of the eco-innovative firms. This documents the prominence of environmental regulation for the initiation of environmental innovations.

Finally we asked some more general questions which are not related to the specific innovation, e.g. concerning the competitive situation of the firm and general employment trends. Again, a differentiation between factors mentioned and the most important factors was made. Price and quality are clearly the dominating competition factors, see Figure 17. By a wide margin, they are followed by innovative products or services, corporate image and environmentally friendly features. Environmentally friendly features are mentioned by only 3% of the firms as the most important factor. This is surprising because the answering firms are exclusively those which introduced environmental innovations in the last three years. It can be concluded that environmental innovations are only developed voluntarily by firms if they face no substantial negative impacts on more important competition factors, especially on costs and quality.

Figure 16: Importance of environmental regulations for processes and products and changes to comply with regulations

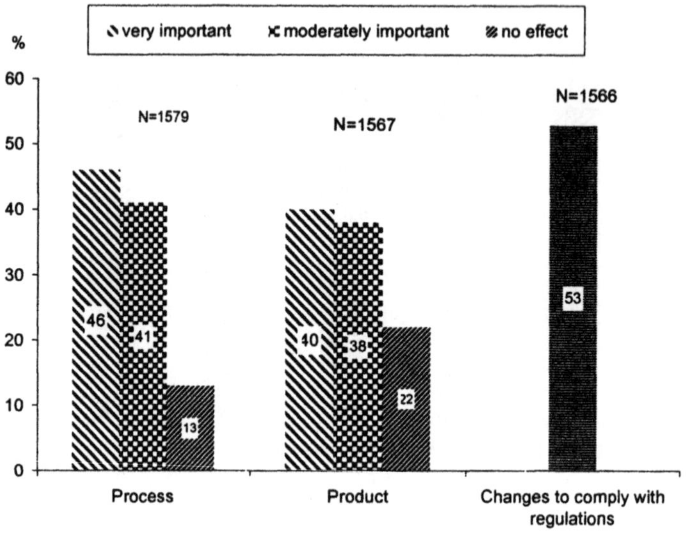

Figure 17: Important competition factors

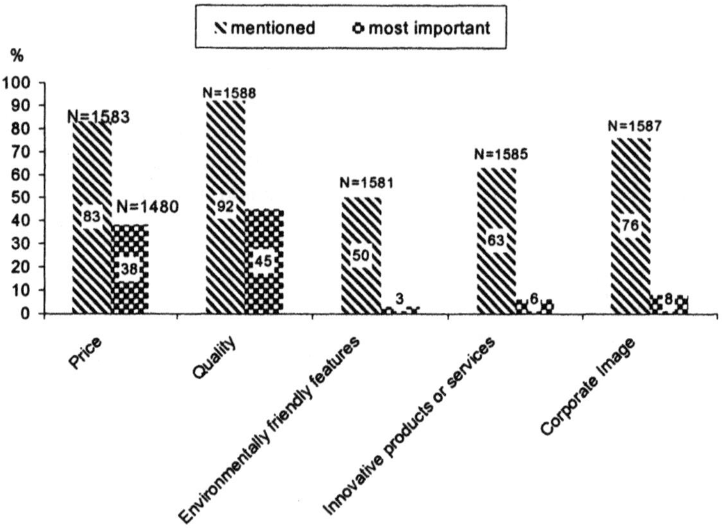

Figure 18 shows the overall employment changes of the firms introducing environmental innovations. For the majority of firms, employment increased in the last three years (49%; see left column). Figure 18 shows that large and small firms increased employment almost to the same extent, while the employment shift to the tertiary sector is represented by a more frequent employment increase in service than in manufacturing firms. The last three columns of the figure present evidence for the measurement error incurred if the general employment change is taken as a proxy for changes in employment induced by innovations. This measurement error only can be avoided in panel studies, see for example Rottmann and Ruschinski (1998). In addition, the difference between total employment change and employment change attributed to the innovation shows that the people who were interviewed were able to differentiate between general employment changes in the firm and changes induced by the eco-innovation.

Figure 18: Change in overall employment

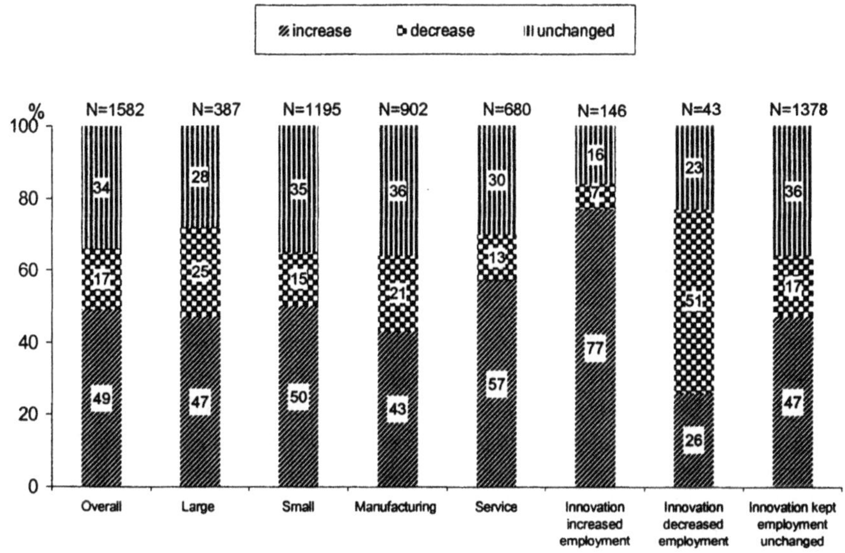

6.3 Interim conclusions

We can conclude from the descriptive analysis that environmentally beneficial innovations generally have a small but positive effect on employment on the firm level. Overall 88% of the firms state that eco-innovations are neutral, 9% state a positive and only 3% a negative impact. For environmental product and service innovations, 18 to 20% state positive effects. These results give first evidence to the hypothesis that more jobs are created by environmentally friendly product and service innovations than by process innovations. Other factors like in-house-development of innovation, size of innovation or regulatory pressure can strengthen both positive and negative effects.

The three most cited reasons for introducing the innovation are to improve the firm's image, to comply with environmental regulation and to reduce costs. This is particularly noticeable for process integrated innovations, recycling innovations and when end-of-pipe (pollution control) technologies were introduced. Increasing market share plays only a minor role for introducing eco-innovations but this reason is particularly important for integrated technologies (product, service and process integrated). On the other hand, price and quality are clearly the most important competition factors for eco-innovating firms. It can be concluded that environmental innovations are only developed voluntarily by firms if they expect no substantial negative impacts on these "hard" competition factors. In the innovation process of a firm, environmental aspects are clearly dominated by economic factors or by restrictions due to regulation. Environment regulation is an important reason for the firms to introduce innovations.

While the data mainly refer to hypotheses on direct effects, at least some general conclusions can be drawn concerning the indirect effects. Income effects due to price changes seem to be negligible, while substantial decreases of energy and waste disposal costs may lead to significant negative indirect employment effects in the energy and waste disposal sector.

7 Econometric analysis of survey data

The goal of the econometric analysis is to quantify the impact of several explanatory variables (such as the impact of the category of innovation, the innovation goals, and the size of the innovation) and suitable control variables on the probability that the eco-innovation has a positive or negative employment effect on the firm level. The dependent variable is the answer to the question "By long-term employment we mean employment in the firm for more than one year. Has this innovation increased, decreased, or had no noticeable effect on the number of long-term employees in your establishment?"

The dependent variable has three values (increase, decrease and unchanged) that are unordered. It can be argued that the determinants of a positive employment impact, of no employment change and of a negative employment impact may be completely different. Therefore, we need an estimation method that allows for differences between the employment changes of firms. We therefore chose a multinomial logit regression, because this estimation technique detects the differences between the determinants for the three values of the dependent variable. The regression thus explains the different reasons for an employment increase or decrease relative to unchanged employment.

The explanatory variables included in the model are directly linked to the considerations above:

- Categories of eco-innovation (see IMPRESS hypotheses)

- eco-innovation goals (since it can be expected that cost reduction targets have a negative impact, while targets to increase market share should have a positive influence on employment) and
- size of innovation (since employment changes only occur when the turn-over costs are more than compensated by the change, see Rottmann and Ruschinski, 1998),
- the firm size,
- whether the effects can be partly explained by subsidies which have been received for the innovation,
- sales expectations (it can be expected that firms with optimistic expectations are more inclined to increase employment already before demand actually increases, this is also called demand pull hypothesis of innovations, see König, 1997 or Rottmann and Ruschinski, 1998),
- whether products or processes have been changed due to environmental regulation (indicator for strictness of environmental regulation),
- competition factors as a proxy for market characteristics (since firms competing on the basis of costs probably choose a different employment behaviour than firms competing on innovative products or environmental performance) and
- the share of workers with unversity or college degree. This may indicate if skills and innovations are complements, i.e. if there is skill biased technological change (see for example Van Reenen, 1997).

The eco-innovation categories, eco-innovation goals, and competition factors are captured by item lists that potentially are correlated and therefore the variables may be collinear. In order to correct for that, a factor analysis was carried out that reduces the number of dimensions to the uncorrelated ones. We use the Kaiser-Meyer-Olkin-Criterion (KMO) in order to decide if the correlation structure of the items is suited for a factor analysis.[2]

For categories of eco-innovation, no clear factors could be identified. The KMO criterion is 0.51 and therefore too low. The highest correlation between two dimensions is only 0.185. In addition, no single factor has a high measure of sampling adequacy and, therefore, multi-collinearity seems to be a minor problem. As a consequence, no reduction in dimensions is carried out here.

For the list of seven innovation goals, the KMO value is 0.67 which allows a factor analysis. There are three components with eigenvalues above 1 which explain 62% of the total variance. The factor loadings can be found in Table 4. The three

---

2 The KMO criterion may lie between 0 and 1. The critical value below which no factor analysis should be carried out is 0.6.

independent innovation goals are given the intuitive names market share, environmental factors and cost reduction.

Table 4: Reasons to introduce the innovation: rotated factor loadings

Reason to introduce innovation	Market share	Environmental factors	Reduce costs	Uniqueness
Comply to environmental regulations	0.29891	**0.78205**	0.08074	0.36754
Secure existing markets	**0.85207**	-0.31464	-0.23076	0.26702
Increase market share	**0.86265**	-0.41575	-0.19023	0.25450
Reduce costs	0.11952	-0.31658	**0.98440**	0.03031
Improve firm´s image	0.49710	**0.59486**	0.10614	0.59457
Respond to a competitor's innovation	**0.54907**	-0.05111	0.09606	0.57204
Achieve an accreditation	0.54134	**0.61711**	0.12153	0.54807

Remark: The bold numbers indicate the factor attributed to the individual innovation reasons.

The competitive situation is captured in a list of five items. Here the KMO value is 0.70 which also allows a factor analysis. There are two components with eigenvalues above 1 which explain 59% of the total variance. The factor loadings can be found in Table 5. Price and quality are subsumed to "hard" competition factors, while environmental and other aspects are called "soft" competition factors. Notice that the first factor is negatively correlated with price and quality and therefore the signs in the regression of this factor have to be reversed in the interpretation of the estimation results.

Table 5: Important competition factors: rotated factor loadings

Competition Factor	"Hard factors"	"Soft factors"	Uniqueness
Price	**-0.83280**	-0.00488	0.30642
Quality	**-0.76436**	0.12803	0.39937
Environmentally friendly features	-0.8364	**0.71760**	0.47806
Innovative products or services	-0.12561	**0.74468**	0.42967
Corporate image	-0.32709	**0.69171**	0.41455

Remark: The bold numbers indicate the factor attributed to the individual competition factor.

Table 6: Regression results of multinomial logit regression

	Prob. that employment increased		Prob. that employment decreased	
	Coeff.	Z	Coeff.	z
Product innovation	.632*	2.349	.536	1.004
Service innovation	.658*	2.205	.432	0.601
Process innovation	-.116	-0.444	.282	0.562
Organisational method innovation	.187	0.548	.051	0.078
Recycling system innovation	-.007	-0.026	-2.361*	-2.180
End-of-pipe innovation	-.055	-0.204	1.593**	3.006
Reason to innovate: Market share	.577**	4.823	.976**	3.801
Reason to innovate: Environment	.008	0.066	-.650**	-2.718
Reason to innovate: Cost reduction	.046	0.385	.692**	2.613
"Hard" competition factors	.094	0.809	-.595	-1.570
"Soft" competition factors	-.105	-0.814	-.503*	-2.119
Large firm	-.832*	-2.369	.666	1.281
Share of environm. on total inno. expenditures	.899**	3.501	1.217*	2.465
Share of total employees with coll/uni. degree	.015**	3.075	.003	0.276
Subsidies or grants for innovation received	.327	1.070	-1.080	-1.205
Sales expectations positive	1.144**	3.191	-.393	-0.751
Change comply with environ. regulations	.753**	2.995	1.345**	2.673
Germany	.941*	2.290	1.231+	1.862
Switzerland	.462	0.907	.737	0.913
The Netherlands	.800+	1.680	-1.185	-1.143
Italy	.404	0.970	-.560	-0.738
Constant	-4.942**	-9.051	-6.269**	-6.496
Number of observations	1,015			
Log likelihood	-353			
Pseudo R^2	0.2253			

Source: IMPRESS Questionnaire, April 2000, Remarks: Significancy levels are as follows: $^{+} < 0.1$, $^{*} < 0.05$, $^{**} < 0.01$.

The bold items are attributed to the respective factors. Note that the first competition factor is negatively correlated with the items price and quality and therefore the signs in the regression also have to be reversed in the interpretation.

The multinomial logit model shows which variables have a stronger impact on firms with a change in employment in comparison to firms without employment changes in the wake of environmentally beneficial innovations. It produces the following significant correlations (see Table 6). A thorough discussion of the estimation results can be found in the last part of this chapter.

A F-test confirms that all parameters are jointly significantly different from zero. The determinants of an employment increase also differ significantly from the determinants of an employment decrease.

Table 7 gives an indication of how well the regression fits reality. First, the predicted probabilities are calculated that one firm is in one of the three different employment situations. Then the firms are classified according to their highest predicted probability. Finally, these predicted outcomes are compared with the actual outcomes. The observations in the main diagonal in Table 8 are predicted correctly with our estimation, while the observations off the main diagonal are not predicted correctly. The table therefore tells us that many firms are predicted to have no employment change although they actually increased or decreased employment. This is a consequence of the weak prediction power of logit models with strongly asymmetrical distributions of ones and zeros. Nevertheless, 88% of the cases have been predicted correctly according to our measure of fit. Table 8 summarises the statistics of the variables used.

Table 7: Measuring goodness of fit: predicted and actual employment changes

		Predicted			
		Increase	Unchanged	Decrease	Total
Actual	Increase	15	90	0	105
	Unchanged	9	874	1	884
	Decrease	0	22	4	26
	Total	24	986	5	1,010

Remark: The 893 observations on the main diagonal are predicted correctly by our empirical model, while the other 117 observations are predicted wrongly.

Table 8: Summary statistics of variables used

Variable	Obs	Mean
Probability that employment increased	1,575	.093
Probability that employment decreased	1,575	.027
Product innovation	1,592	.174
Service innovation	1,591	.118
Distribution system innovation (reference)	1,592	.085
Process innovation	1,592	.363
Organisational method innovation	1,592	.129
Recycling system innovation	1,592	.318
End-of-pipe innovation	1,592	.318
What percentage of all employees has a college or university degree?	1,321	18.72
Innovation share larger than 50%	1,284	.181
Large firm	1,594	.245
Subsidies or grants for innovation received	1,521	.105
Sales expectations positive	1,482	.789
Change to comply with environmental regulations	1,566	.534
UK (reference)	1,594	.251
Germany	1,594	.252
Switzerland	1,594	.126
The Netherlands	1,594	.130
Italy	1,594	.241
Number of firms answering all questions (net sample)	1,015	

8 Conclusions and policy implications

There are different methods for analysing the employment effects of environmental innovations. There is no single best method as all methods have their own particular shortcomings. It is difficult to ascertain the different types of employment effects of innovations, or to establish the overall effects. For analysing the employment effects of innovation the following methods may be used: survey questionnaires (mail surveys, telephone questionnaires), case study analyses of innovating companies, product chain analyses, input-output analyses and general equilibrium models. Table 9 describes the kind of employment effects that may be analysed and estimated through the various methods.

8.1 Conclusions from case studies synthesis analysis

While partly related to the assessment of employment effects, the case study phase of the project was largely designed as an input into the development and refinement of a methodology for analysing these employment effects.

Before summarizing the findings, it is important to stress that the main result of the case studies is that employment effects are distributed along the value chain and occur only partly at the level of the firm adopting the eco-innovation.

8.1.1 Direct vs. indirect employment effect

Case studies provide an interesting insight on the different direct and indirect employment effects along the value chains considered.

The case studies aimed at addressing both direct and indirect effects, but indirect income effects were only partly considered due to the difficulties in gathering information in a consistent manner.

In addition, the apparently clear distinction between direct and indirect effects has turned out to be difficult to implement in some specific circumstances. In this respect, the case of eco-innovative financial services is very illustrative, showing the difficulty in defining the value chain and the subsequent identification of direct and indirect effects. In this case, the upstream part of the value chain is made of savers that support the loan making activity.

The results demonstrate that the direct employment effect at the level of the adopting firm is normally very small or negligible, with a slight exception in the cases where the adopter of the innovations is also the developer of it. In these cases, like in some examples of biotechnology, the research and development activity can generate new jobs and require new skills, but the effect is likely to be a temporary one.

Table 9: Methods for analysing the employment effects of (eco)-innovation

Type of method	**Type of employment effect to be analysed**	**Strong and weak points**
Econometric analysis of company (plant) data on innovative behaviour and employment	Overall intra-company effects	Possible to take account of additional control variables and assess significance of effects. Not possible to separate direct and indirect effects at company level or to link employment effects to specific innovations.
Survey questionnaire of innovating companies	Direct effects and indirect substitution effects of specific innovations in the innovating companies	Possible to generalise the direct and indirect substitution effects of different types of innovation (integral process change, end-of-pipe, recycling etc.). Not possible to analyse employment effects that occur outside the innovating company.
Case study analysis of companies adopting or developing an eco- innovation	As above for now for company as a whole or for company unit.	Possible to get a better understanding of the processes through which employment and eco-innovation are linked within the companies studied. Temporal (adjustment) may be studied. No generalisation possible about employment effects from eco-innovation.
Production chain analysis	Direct and indirect substitution effects in the chain producing and using the innovation	Possible to analyse the effects across a chain. Not possible to analyse compensatory effects that occur through income changes that stem from changes in value added.

Table 9: continued

Input-output analysis	Employment changes from an innovation throughout the production chain and supplying and using sectors	Employment effects may be established for every sector that is in some way connected to an innovating industry. Income effects from the innovation and price effects may be incorporated in the analysis. Labour intensity of the innovation may differ from the average labour intensity of the sector. It is difficult to break down the employment effects.
General equilibrium modelling	Overall effects for the economy as a whole	Possible to take into account income or compensatory effects. Not possible to estimate direct and indirect effects with any precision. Estimates of overall effects sensitive to modelling assumptions.

8.1.2 Short term vs. long term employment effect

The issue of time frame for assessing the employment impact of eco-innovations is obviously very relevant since many innovations need time to be accepted by clients and customers and since future environmental policy can create the ground for some of these innovations to succeed in the marketplace.

The difficulties are very clear for example in the case of recycling of CFC-refrigerators, where the company involved in the recycling activity reported tremendous changes in the labour force.

In this case the point of view of the researcher is retrospective and, theoretically, the author could rely on information going back ten years to understand determinants and effects of employment changes. However, in practice a number of events and factors (such as change in market structure, changes in legislation, changes in prices of materials and secondary raw materials, changes in the internal organisation of processes) have clearly diminished the possibility for the researcher to determine the employment effects of the innovation in a dynamic situation.

The case study of biomass energy production and distribution is very instructive, by underlining the same difficulties mentioned above and, at the same time, introducing new dimensions to the problem. In this example the introduction of the innovation is relatively new and the point of view of the researcher is retrospective and prospective at the same time.

Practical problems relate to more conceptual difficulties in understanding what are the system boundaries along the axis of time: How far should the analysis go back in considering for example the phase of construction of the plant and related pipelines? How far should the analysis go forward in considering the dismantling phase and substitution of the plant?

A third interesting problematic situation refers to the biotechnologies where the adoption of some of the innovations is new and still marginal when compared to the potential reference market. In this case the researcher needs to project results into a situation where the innovation is widespread, taking into account the way in which the diffusion of the innovation could effect involved firms.

8.1.3 Local vs. regional employment effect

To a small extent certain case studies address the issue of geographical distribution of employment effects, which can be a very relevant aspect for policy makers aiming at promoting employment possibilities in some specific regions of Europe.

This is addressed, for example, in the case of biotechnologies for starch conversions into sweeteners, where enzymes are used, instead of chemicals. While the employment effect at the level of natural resources and extracting firms is neutral for Europe, it has substantial positive employment effects when the adopter of the innovation is based in the US. In this example, maize is the input material for sweeteners instead of imported sugar and the large maize availability in the US makes the innovation more employment friendly.

A second example concerning the issue of geographical distribution of employment effects is the energy production from the biomass example. Here the geographical scale of employment effects is very clear. When shifting the fuel from non renewable oil based sources to biomass from the wood industry and local agriculture there is clearly a shift of employment from global to local. Nevertheless, the consideration of a wide value chain makes the evaluation more complex since other aspects (e.g. connected with the plant technology) work in the opposite direction.

The issue of local sources also plays an interesting role (in terms of employment) associated to transportation: the use of local sources like biomass, would require less travel, less environmental impact but also a smaller employment effect on a wider scale.

8.1.4 Generalisation of results and lessons learnt

This brief discussion of key findings of the case studies has underlined the practical and conceptual difficulties that make each case in some way unique.

Nevertheless it is here important to address the questions:

- does the value chain approach make sense in providing an insight into the employment effect?

- are the results general? And therefore, what are the lessons that a policy maker can derive from the use of such an approach?

The first question has been implicitly addressed several times in this document; the case studies suggest that the direct effects at the level of the innovating organisation are very limited (both positive and negative) and the judgement about the employment-friendliness of all the innovations studied cannot be made without taking into account what happens outside the adopting firm. The value chain approach introduces a wider perspective by considering the upstream and downstream activities where most employment effects can occur.

On the other side it is clear that the value chain approach is not able to capture some of the employment effects determined by the innovation. The limitation is related to the vertical approach adopted in the methodology which makes it difficult to understand the horizontal effects on competitors (they are an important component of the compensatory effect). A possible way to handle this problem would be to consider the competitive environment where the innovation has taken place, trying to understand if the relative weight of competitors (with respect to market share and sales) has changed in relation to the innovation. This has been done effectively in the case of energy efficient windows where the growth trends of the innovating company have been compared with those of the construction sector as a whole.

On the other hand, while this could be done relatively easily in the case of innovations on product and services, it can be an impossible task for process innovations, organisational measures, and end-of-pipe innovations. In these cases it can be much more arbitrary to find relationships between the adoption of the innovation itself and the improved or reduced strength of the business in the marketplace.

The second question introduces a more challenging issue related to the objectivity of results. The history of LCA provides, once again, some help. LCAs have been used in the early stage for advertising and public relations purposes in order to demonstrate that one product was better, from an environmental point of view, than some alternatives. Later, this kind of misuse of LCA has been abandoned due to lack of consistency of results and the possibility, for virtually any company, to demonstrate the opposite by redefining the system boundaries and changing algorithms for the calculation of the environmental impact.

However, not only methodological lessons can be learned from the case studies. Concerning the IMPRESS hypotheses, some first qualitative insights can be drawn from the LCA analysis. Firstly it seems, as expected, that environmental product innovations (e.g. energy-saving windows) create more jobs than process innovations (use of enzymes). Secondly, environmental regulation has been identified as a major determinant for employment since labour demand depends on environmental standards. Similar to other technological trajectories, labour-saving technological progress can be observed for environmental technologies in the long run, too. These first qualitative insights were used as a basis for getting further, more quantitative empirical evidence concerning our hypotheses. This was done by the IMPRESS telephone survey.

8.2 Conclusions for policy implications from survey synthesis analysis

In this section we discuss the factors which have been identified to be significant for an increasing probability of positive employment changes, for negative employment changes, or for both. Finally we draw some policy conclusions.

Firstly, several factors are significant for positive employment impacts of environmentally beneficial innovations. In comparison to the reference category (logistics innovations[3]), product and service innovations have a positive effect, while all other innovation categories are not significant. This is in line with the literature on innovation and employment. Indirect substitution effects of product and service innovations on the firm level appeared to be lower than we expected. To control for this substitution effect, the question "Did this innovation replace some of the product/service sales of your establishment?" was used. While we expected substitution effects in nearly all cases, only 43% of the product innovators and 27% of the service innovators answered "yes". It seems that ecological products and services have created their own market niches and are supplements to conventional goods and services. However, it can be assumed that for most of the innovations mentioned by the firms, as for example "new cleaning techniques" or "transport reduction measures", negative indirect substitution effects occur in other firms.

Small firms report more employment increases than large firms. This result is in line with other empirical studies on the general relation between innovation and employment, too (see Brouwer et al., 1993 or Smolny and Schneeweis, 1999). Firms with high shares of employees with college or university degrees have a higher probability to increase employment in the wake of innovations. This may be an indication that environmentally-oriented innovations are skill-biased, see Van Reenen (1997) and Chennels and Van Reenen (1999). Positive sales expectations are highly significant for increasing employment (this is also found in most other estimations on the firm level, see the discussion in König, 1997). As innovations frequently are induced by positive sales expectations, there is a possible endogeneity problem here, see Chennels and Van Reenen (1999). Firms in Germany, the Netherlands and Italy have a higher probability to report a positive employment effect than firms in the UK, the reference country.

---

3 Innovations in logistics were used as reference category because they have a small share of the total sample and appeared to be not significant in the econometric analysis.

Secondly, different factors have been identified to correlate significantly with an increasing probability of job losses. End-of-pipe processes have a positive, recycling innovations a negative correlation with the probability that the innovation has a negative impact on employment. This result is quite surprising since we expected the same direction of employment effects for both kinds of innovations. However, already the descriptive analysis shows that employment effects of recycling innovations are positive in almost all cases. An explanation for the difference between end-of-pipe technologies and recycling measures may be the maturity of regulation. While end-of-pipe regulations have existed, in many cases, for twenty or thirty years, political measures concerning recycling were mainly implemented during the 1990s. Thus, new end-of-pipe innovations often replace existing older technologies, while other environmental process innovations, especially in the area of recycling, have lower substitution effects. We have addressed this substitution effect by the following two questions in the questionnaire "Did the introduction of this process innovation replace, at least in part, a previous production process in your establishment?" and "Did the introduction of this process innovation replace, at least in part, end-of-pipe pollution control equipment?" 38% of end-of-pipe innovators answered the first question with "yes", compared to 19% of recycling innovators. And 34% of end-of-pipe innovators agreed on the second question, compared to 13% of recycling innovators. These numbers indicate that substitution effects are substantially different across different types of environmental process innovations. This may explain the better performance of recycling innovations in terms of employment.

When environmental goals motivate the innovation, it is less probable that the innovation has a negative employment effect. On the other hand, cost reduction as motivation for the innovation increases the probability that the firm reduces employment. It can be expected that firms invest in improved labour-saving technologies, especially in areas where the technologies have already reached a certain maturity (mainly end-of-pipe technologies). If the firm is not under the pressure of strong cost competition, "soft" factors like environmental aspects become more important. This decreases the probability of job losses.

Furthermore, three factors have been identified which can affect employment in both directions. Market share as an innovation goal, size of innovation and strictness of environmental regulation are significant for either positive or negative employment changes. Market-oriented strategies focusing on the development of environmentally beneficial innovations impose chances but also risks on firms. Environmental products and services are often marketed on small niches and thus bear risks of profitability. Other innovation goals like complying with environmental regulations still dominate. Concerning innovation size, the result confirms the hypothesis that costs to hire or fire employees lead to changes in employment only if there is a major re-organization in the wake of the innovation. A critical innovation size must be reached before employment changes are measurable in person/years by surveys as carried out within our project. Also environmental regulation can lead to both positive or negative employment effects, depending on the concrete innovation actitivities being undertaken due to the regulatory pressure. Regulations, for example that introduce a new step into production like waste treatment can increase employment within a firm while regulations that force firms to increase the capital intensity of production can reduce employment.

Finally, it is remarkable that subsidies or grants for the innovation do not have any employment impact. This means that the allocation of subsidies for environmental technologies and innovations in the respective countries is neutral concerning employment and thus does not counteract labour market policy. However, employment could be stimulated by shifting more money from end-of-pipe measures to integrated measures, especially products and services. But even an employment-oriented allocation of subsidies for environmental programmes will only induce minor employment changes.

To sum up the results of the econometric analysis, basic hypotheses about the general relationship between innovation and employment have been confirmed. Product and service innovations create more jobs than process innovations. Employment changes only occur in the wake of major innovations. This is confirmed by the significant impact of the share of innovation expenditures on employment increases as well as employment decreases. We also detect signs of skill-biased technological change of eco-innovations, because the share of highly qualified employees has a positive impact on employment increases, while it is insignificant for employment decreases. Significant impacts of the control variable size show that economic innovations create employment especially in small firms. A positive correlation between sales expectations and employment is in broad conformity with the innovation literature.

Beyond these general insights, some specific conclusions can bc drawn for a coordinated environmental, innovation and labour market policy. Generally, eco-innovations have a small but positive effect on employment on the firm level. Thus environmental support programmes do not counteract labour market policy. However it should not be expected that ecological modernisation of industries contributes substantially to a reduction in unemployment. A further shift from end-of-pipe technologies to cleaner production, especially towards product and service innovations, would be beneficial for the environment and would create jobs. This synergy should be considered in political programmes whether they are motivated by environmental, labour market or innovation policy. Some potential still exists for shifting subsidies from end-of-pipe solutions to cleaner products and services.

The scope of the data is limited to employment effects on the firm level when a firm classified itself as being an eco-innovator. Improving the knowledge about either the universe of eco-innovators or the differences between eco-innovating firms and other firms are important questions for further research. While the problem of measuring innovations is addressed thoroughly, we cannot control for fixed effects and possible endogeneity for example in demand expectations. While the data mainly refer to hypotheses on direct effects, at least some general conclusions can be drawn concerning indirect effects. Income effects due to price changes seem to be negligible, while indirect substitution effects of product and service innovations seem to occur mainly beyond the firm level. Indirect substitution effects of process innovations seem to depend on the maturity of regulation, i.e. new regulation requires new equipment while older regulation means that equipment already exists and is replaced in the innovation process.

References

Bovenberg A. L. and F. Van der Ploeg (1994), *Consequences of Environmental Tax Reform for Involuntary Unemployment and Welfare,* Center for Economic Research, Discussion Paper No. 94/08.

Brouwer, E., A. Kleinknecht and J. Reijnen (1993), Employment Growth and Innovation at the Firm Level – An Empirical Study, *Evolutionary Economics* 3, 153-159.

Chennels, L. and J. Van Reenen (1999), *Technical Change and the Structure of Employment and Wages: A Survey of the Micro-Economic Evidence*, Working Paper 99/23, Institute for Fiscal Studies, London.

Cleff, T. and K. Rennings (1999), Determinants of Environmental Product and Process Innovation – Evidence from the Mannheim Innovation Panel and a Follow-Up Telephone Survey. in: H. Karl and C. Orwat (Eds.), *European Environment, Special issue on Integrated Product Policy, Vol. 9, No. 5*, 191-201.

Coenen, R. et al. (1995), *TA-Projekt Umwelttechnik und wirtschaftliche Entwicklung: Integrierte Umwelttechnik Chancen erkennen und nutzen*, TAB (Büro für Technikfolgenabschätzung beim Deutschen Bundestag), Arbeitsbericht 35, Karlsruhe.

Conrad, K. and T. Schmidt (1997), *Double Dividend of Climate Protection and the Role of International Policy Coordination in the EU – An Applied General Equilibrium Analysis with the GEM-E3 Model*, ZEW Discussion Paper No. 97-26.

European Commission (1993), *White Paper on "Growth, Competitiveness and Employment" No. 145* , Brussels.

European Commission (1997), *Communication on Environment and Employment.* Brussels.

European Environmental Agency (1997), *Life Cycle Assessment (LCA): A Guide to Approaches, Experiences and Information Sources*, Copenhagen.

Fuzzler, C. (1996), *Driving Eco-Innovation*. London.

Gameson et al (1997), *Environment and Employment – A report prepared by the IPTS for the Committee on Environment, Public Health and Consumer Protection of the European Parliament,* Sevilla, Spain.

Goodstein E. (1995), Jobs or the Environment? No Trade-Off, *Challenge* No. 01-02, 41-45.

König, H. (1997), Innovation und Beschäftigung, in: Vosgerau, H. and H. König (Eds.), *Zentrum und Peripherie – Zur Entwicklung der Arbeitsteilung in Europa*, Schriften des Vereins für Socialpolitik, Vol. 250, Berlin, 149-176.

König, H., H. Buscher and G. Licht (1995), Investment, Employment and Innovation, in: OECD (Ed.), *Investment, Productivity and Innovation*, Paris, 67-84.

Köppl, A. and C. Pichl (1997), *Wettbewerbsvorteile durch umweltorientierte Innovationen. Überprüfung der First-Mover-These,* Umwelt und Arbeit II, Informationen zur Umweltpolitik Nr. 122 der Arbeitskammer Wien.

Kosz, M. (1997), *Integrierter Umweltschutz und Arbeit. Erste Erfahrungen und langfristige Perspektiven*, Umwelt und Arbeit III. Informationen zur Umweltpolitik Nr. 123, Institut für Wirtschaft und Umwelt der Arbeitskammer Wien.

Pfeiffer, F. (1999), Human Capital and Innovation in Eastern and Western Germany, in: M. Fritsch and H. Brezinski (Eds.), *Innovation and Technological Change in Eastern Europe*, Edward Elgar, Cheltenham, 142-166.

Pfeiffer, F. and K. Rennings (2001), Employment Impacts of Cleaner Production – Evidence from a German Study Using Case Studies and Surveys. *Business Strategy and the Environment*, Vol. 10, No. 3, 161-175.

Porter, M.E. and C. van der Linde (1995), Green and Competitive: Ending the Stalemate, *Harvard Business Review* Vol. 73, No. 5, 120-134.

Ritt, T., H. Mahringer, M. Getzner and O. Fitz (2000), *Umwelt und Beschäftigung: Strategien für eine nachhaltige Entwicklung und deren Auswirkungen auf die Beschäftigung*, Institut für Wirtschaft und Umwelt der Arbeitskammer Wien, Vienna.

Ritt, T. (1999), Integrierter Umweltschutz: Enttäuschend gut! *UmweltWirtschaftsForum*, Vol. 7, No. 2, 24-28.

Rottmann, H. and M. Ruschinski (1998), The Labour Demand and thc Innovation Behaviour of Firms – An Empirical Investigation for West German Manufacturing Firms, *Jahrbücher für Nationalökonomie und Statistik* Vol. 217, No. 6, 741-752.

Smolny, W. and T. Schneeweis (1999), Innovation, Wachstum und Beschäftigung – Eine empirische Untersuchung auf der Basis des Ifo Unternehmenspanels, *Jahrbücher für Nationalökonomie und Statistik* Vol. 218, No. 3+4, 453-472.

Van Reenen, J. (1997), Employment and Technological Innovation: Evidence from U.K. Manufacturing Firms, *Journal of Labor Economics* Vol. 15, 255-284.

Vergragt, P. J. and L. Jansen (1993), *Sustainable Technological Development: the Making of a Dutch Long-Term Oriented Technology Programme,* Project Appraisal, Vol. 8, No. 3, 134-140.

Weizsäcker, E. U. von, A. B. Lovins and H. L. Lovins (1995), *Faktor 4 – Doppelter Wohlstand, halbierter Naturverbrauch*, München.

Part II: Case Studies

Windows

Mathias Binswanger and Daniel Schulthess

1 Introduction

The effects on employment of eco-innovations are examined using windows as a case study. Depending on the life cycle of the item under investigation, the employment effects during the various stages of production, installation, repair and disposal of a window are to be examined.

To this end, an initial section explains the role of the Swiss construction industry against the background of its relevance to employment. In order to place the window as an eco-innovation in its ecological context, in a second stage the environmental relevance of the construction industry, key ecological problems and ecological demand are analysed according to their salient features.

Finally, in the main part of the work, the employment effects of the window sector – as part of the construction industry – are discussed and clarified by means of analysing the impact on employment at two Swiss window manufacturers.

2 Description of the construction industry

2.1 The role of the construction industry in the Swiss economy

The construction industry is of great importance to the Swiss economy. In 1995, about 10% of the total labour force was engaged in the construction industry, which also accounts for about 7% of the Swiss GDP. This percentage is higher than in any other Western European country. Moreover, changes in construction activities have a large impact on the Swiss economy and employment, as most intermediate products used in construction are from domestic producers. Consequently, almost all of the positive direct or indirect employment effects that may be caused by innovations will be felt within the domestic economy, where they are much easier to track than abroad.

The development of the construction industry since World War II has been very dynamic. Until 1990 construction activities grew faster than the total Swiss economy as there was high demand from the private sector as well as the public sector.

But this trend was completely reversed in the early 1990s. While total spending for construction activities increased from about CHF 20 billion in 1975 to about CHF 50 billion in 1990, it declined again to about CHF 40 billion in 1998. This decline in construction activity also led to a substantial decline in employment. In the construction industry employment decreased by more than 11% between 1991 and 1995 – the largest reduction of all industries. A closer look at this development shows that the shrinking of construction activities was caused by the heavy reduction in the construction of new buildings (especially in the private sector), which declined by about 30% from 1990 to 1998. The renovation of existing buildings, on the other hand, increased slightly over the same period. Consequently, there was not only a major decline in the overall level of construction activity but also a shift from the construction of new buildings towards the renovation of existing buildings. This shift also created additional demand for products that are used in the upgrading of existing buildings such as windows with better insulation features.

Compared to other industries, the construction industry is characterised by a very low level of concentration. Most of the firms are small or medium-sized and an average firm has about ten employees. Two thirds of all firms have less than 50 employees and 90% have less than 200 employees. Less than two percent of all firms have more than 500 employees.

The construction industry may be divided into four sub-sectors, which are design (architecture, engineering), main construction activities (above-ground construction, foundation engineering), interior working (installation of appliances) and the supporting industry (SBV, 1993). Main construction activities and interior working each account for about one third of the total employment related to the construction industry. If we use these broader definitions of the construction industry, over 500,000 people are working in jobs that are related to construction activities.

2.2 Ecological analysis of the structure of the industry[1]

In order to understand the ecological problems of the window industry in their entirety, it is essential on the one hand to look further than actual window production and to extend one's horizon to include the upstream and downstream stages. On the other hand, since windows constitute an inseparable component of buildings, in an initial stage the fundamental ecological problem areas must be indicated, considering the life cycle of buildings. The product life cycle of a window is therefore illustrated within the framework of concrete case studies.

1 The following explanations are supported by an investigation of the "Ecological Change in Swiss Industries" (Dyllick et al., 1994).

Figure 1: Environmental impact of different industry stages

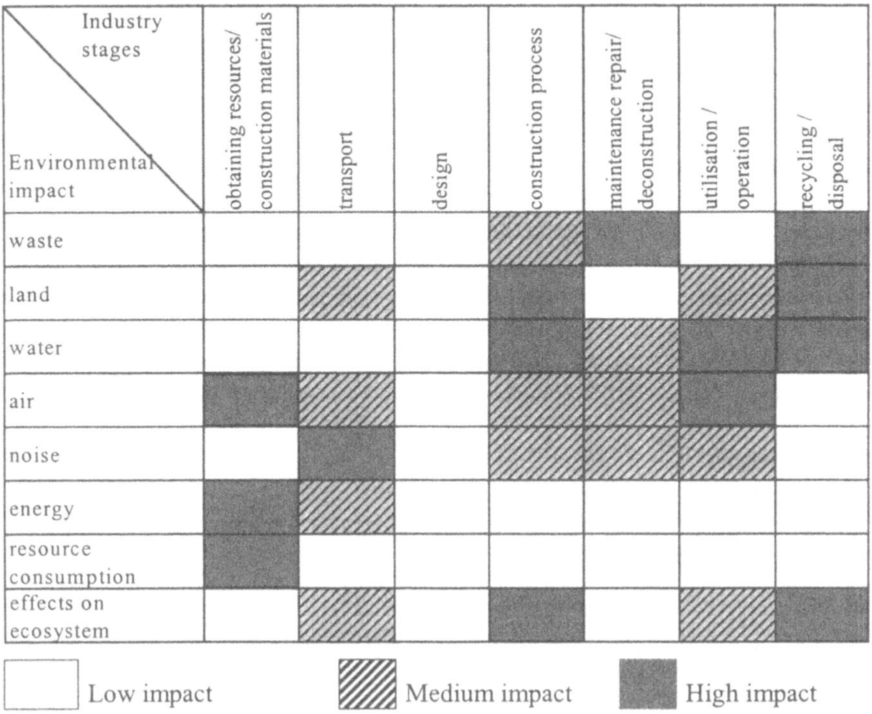

The industry can be roughly subdivided into seven stages: obtaining raw materials and producing construction materials, transport, design, the construction process, repair and demolition, utilisation and operation, recycling and disposal. The environmental impact of the individual industrial stages are assessed using the parameters of waste, land, water, air, noise, energy, consumption of resources and effects on ecosystems. Figure 1 shows the environmental impact generated at the individual industrial stages.[2]

2.3 Obtaining raw materials / manufacturing the product

In the first stage of the industry, the main sources of environmental impact are primarily the high energy requirement for the manufacture of construction products and the associated emissions of air pollutants (especially CO_2, SO_2 and NOx), as well as a general reduction in mineral reserves. For the window industry, this is primarily the production of aluminium and PVC for the window frames[3]. The

---

2 Koller, 1994: 190.

3 The assessment of raw materials is based on the results of the investigation "Umweltbewusste Bauteil- und Baustoffauswahl" (Environmentally aware choice of construc-

various chemical processes involved in the production of raw materials require particularly critical assessment from the ecological viewpoint. The toxicity of the initial, semi-finished and end products, as well as the by products of the product stages, constitute a considerable hazard potential with regard to pollution of the soil, water and air.

2.3.1 Transport

Approximately 33% of the goods transported by road are construction materials. The construction industry is therefore responsible for about one third of all transport which takes place in Switzerland.[4] Energy consumption, atmospheric pollution and above all noise pollution generated by the construction industry are therefore not to be underestimated. Nor must one neglect the fact that transport always involves use of land. Traffic is the second largest consumer of land in Switzerland.

2.3.2 Design

The direct environmental impact due to the design sector of the industry is generally classified as low.

Indirectly, design predetermines the ecological effects throughout the life cycle of the product. As a result, the participants in the design phase are well placed to reduce the total environmental impact as effectively and efficiently as possible.

2.3.3 The construction process

Actual building activity itself has onerous effects on the environment; the most important are:

- The effects on ground water, such as lowering of the water table[5], diversion of ground water flows and contamination.
- Land use. Every year in Switzerland some 50,000 construction sites use approximately 32 mln m^2 of building land.[6]
- In addition, during construction various "waste products" are produced, such as construction plant exhaust gases, building waste and noise; some of these have a substantial impact on the environment.

---

tion components and materials) by the Ministry for Building and Housing of North Rhine-Westphalia (MBW) (Geissler et al., 1993).

4 Hugenschmidt, 1993: 21, Bundesamt für Statistik, 1986.

5 Gesundheits- und Wirtschaftsamt der Stadt Zürich, 1992.

6 See Wüest and Gabathuler, 1990: 20-21, Köppel, Schmitt and Leiser, 1992: 90. This corresponds to a building rate of approx. 90,000 m^2 or 8 football fields per day.

2.3.4 Repair / demolition

With the shift in building activity from new building to conversion and renovation, the repair and demolition phase is becoming increasingly important. At 3.9 mln m^3 per year, waste constitutes the most serious environmental impact in this phase.

It is not only the amount but also the physical and chemical composition of waste which is problematic from an ecological viewpoint. Moreover, conversion work is also associated with noise pollution.

2.3.5 Utilisation / operation

If one considers the product life cycle of a building in its entirety, the impact resulting from use, in terms of materials and energy, is extremely high; energy consumption (via heating systems), water use (consumption of drinking water) and air pollution (heating again) constitute the main areas which have environmental impacts.

With particular reference to the "window" case studies it is important to know that today ten times as much energy is consumed in the utilisation phase of a building as during its construction. Energy consumption due to heating systems is very significant and is associated with a correspondingly high contribution to air pollution – especially in terms of greenhouse gases such as CO_2 . In this context, energy-saving windows can make a substantial contribution to reducing energy consumption.[7]

2.3.6 Recycling / disposal

Closing the circle of materials use will become more and more critical in the future. In the case of buildings there are numerous applications where prepared building rubble can be recycled without problem, but recycling often represents a danger to the soil, water, flora and fauna because of the sometimes environmentally hazardous composition of the unprocessed materials and the ever-increasing concentration of pollutants in the construction materials which have to be disposed of.[8]

The window industry is attempting to close the circle of materials use. As indicated in the raw materials/construction materials production stage, with the ex-

---

7 Cf. also IP Bau/Bundesamt für Konjunkturfragen, 1992: 19, Keller, 1992: 45.

8 Cf. Steiger (1991), who predicts that construction materials dating from the 60s and 70s will be difficult to dispose of. Since to date the majority of demolished buildings are older than this, it is to be expected that the clearance and demolition of newer structures will distinctly increase the amounts of problematic waste, which is difficult if not impossible to separate and recycle.

ception of glass, recycling technology is most advanced in the case of aluminium. Even with plastic frame materials, more attention is being paid to ease separation and recyclability, although this has only occurred over the last few years.

Final storage of the non-recyclable construction materials takes place in so-called residual materials dumps or reactor dumps. The chemical and biological reactions within these dumps constitute potential "legacy hazards"; in the short term, their emissions may well cause locally limited soil contamination, as well as ground water and surface water contamination, but in the medium and long term they may also contaminate ground water at the regional level.[9] This emission potential of dumps can have a very large impact on local and regional ecosystems.

The ecologically problematic industrial stages therefore include recycling and disposal, utilisation and operation, and the actual construction process. These three phases therefore constitute the most important starting points if environmental impacts in the industry are to be reduced.

Figure 2 below illustrates the key ecological problems of the construction industry, with key ecological problems of the window industry shown in bold type.

2.4 Key ecological demands

The preceding chapter considered ecological problems from an energy/materials viewpoint; this chapter focuses on pressure groups and control systems. For companies and industries, environmental problems are relevant only if and to the extent that they impinge on the company as demands from market, political and social actors.[10] According to the control system concept, pressure groups are institutional representatives of three different control systems: the market, politics and the public.[11]

As an entry point into an analysis of the effects of the market, politics and the public on the construction industry in general, the table below contrasts the industrial stages of the control systems (see Figure 3).

---

9 Baccini, Belevi and Lichtensteiger, 1991: 2.

10 Dyllick, 1992.

11 On the control system concept, see Dyllick et al., 1994.

Figure 2: Key ecological problems of the construction industry

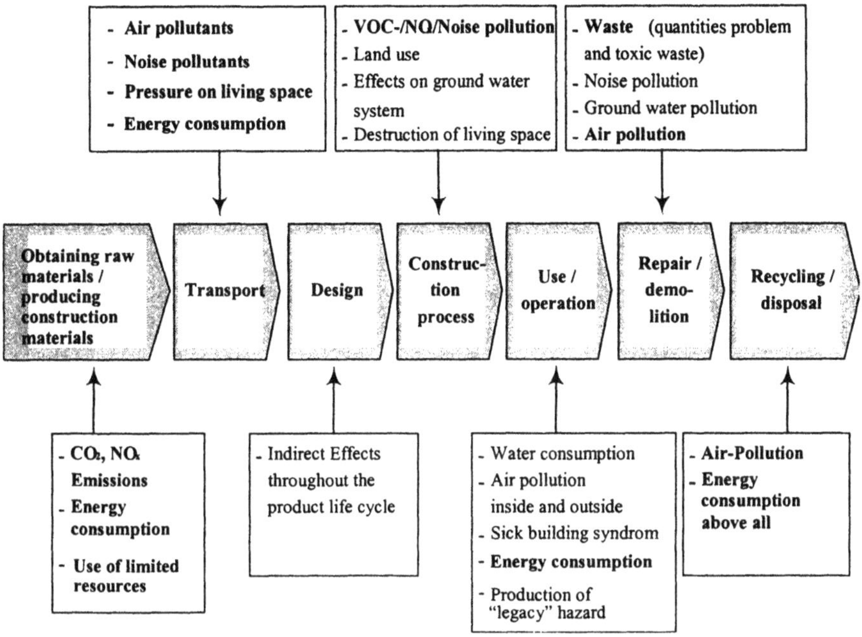

Figure 3: Impact of pressure groups during industry stages

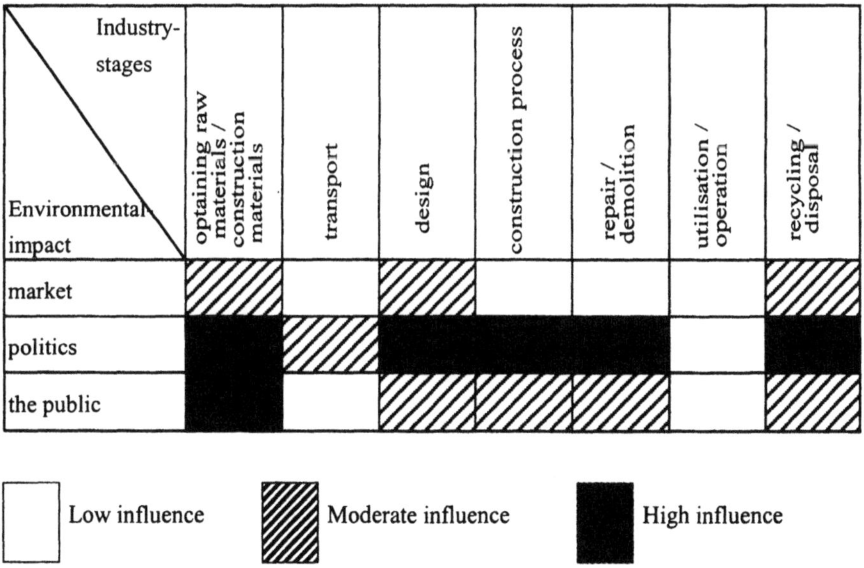

2.4.1 Control system: the market

The following pressure groups are deemed relevant in the market control system: financiers, insurance companies, construction materials suppliers, architects and engineers, designers, disposal companies and recycling companies, plus their employees and clients.[12]

With regard to ecological demands, the relevant pressure groups in the construction industry are the materials suppliers, designers and finally the clients.

On the one hand, the pricing policies of construction materials suppliers affect decisions on the use of materials, energy reduction, use of recycled materials and substitution in the narrow sense, i.e. the replacement of conventional building materials by ecological products. On the other hand, construction materials suppliers influence designers, contractors and clients via their use of advertising campaigns. Here, in addition to functionality, safety, reliability and weather resistance, the ecological properties of products are becoming increasingly important. This trend can be observed particularly in window construction, where in addition to thermal insulation, the recyclability of components and the use of environmentally friendly paints, lacquers and adhesives are being stressed as important product characteristics.

Through their advice and specifications, designers have a substantial influence on the behaviour of clients, the use of materials, the performance of the building and the development company and thereby on the recyclability of the building and its components. The values espoused by designers are therefore very important for a more ecologically sound method of construction. However, they must also be provided with the necessary information, since it is not easy to maintain an overview when there are tens of thousands of construction materials and components.

The final decision on the use of materials and components lies with the client, and the ecological standards and values of the owners of the building play a central role. In many cases, financial considerations are still an obstacle to ecologically-oriented decisions and behaviour.

2.4.2 Control system: politics

In addition to defending a market, the legitimisation of company behaviour with regard to political bodies is a central prerequisite for guaranteeing the continued existence of a company. The political control system, thanks to the threat of applying sanctions, can have a rapid and decisive effect on company behaviour.

From an ecological point of view, the legitimisation of behaviour with respect to the political control system is ensured by complying with legal environmental

---

12 Koller, 1994.

regulations, official conditions and voluntary agreements, not to mention participation in the further development of the political parameters.

The construction industry in general and the window industry in particular are affected very directly by legislation on environmental protection:

- Regulations on maintaining air purity,
- Technical ordinances on waste,
- SIA standards at canton level,
- Particularly for the window industry, the voluntary code of practice of the trade association (Fachverband Fenster- und Fassadenbaubranche, FFF),

but it is also affected indirectly, e.g. in the case of raw materials production and processing, by the law on water protection (GschG). An indirect but major influence is also exercised by planning laws in conjunction with building regulations at the canton level. These directly affect thermal insulation – of central importance in window construction – and the concrete specifications for materials and construction work.

Laws are implemented by authorities. However, this pressure group is very difficult to evaluate in terms of its impact, firstly because substantial differences in implementation are apparent within and between cantons; these are shown in divergent environmental costs and conditions.[13] Secondly, because the authorities are highly dependent in terms of the application of the law on public opinion and the situation regarding environmental protection.

2.4.2.1 Control system: the public

The nature of environmental effects implies that they only partially occur at the company's location. Apart from politics, therefore, the 'public' control system plays an important role in legitimising commercial activity. The public is represented by the following pressure groups: the media, environmental protection organisations, residents' associations and interest groups, plus scientific institutions.[14]

As an infrastructure and a medium, the media occupy a central position in the formation of public opinion. The impact of the media is detectable at all the stages of the industry, but in particular in design and in the production of construction materials.

The environmental protection organisations are one important pressure group in the area of environmental concerns. The building permit procedure and above all the environmental compatibility audit give such organisations the right in law to

---

13 Schwank, Koch and Mauch, 1992.

14 Dyllick et al., 1994.

lodge a complaint as an association.[15] However, the production of construction materials has already been the target of actions by environmental protection organisations. The case of the construction material asbestos demonstrated this very effectively.

The dependence between implementation by authorities and the public has already been mentioned. Here is it precisely the residents themselves who play an important role as a result of their direct physical presence. Production, recycling and disposal installations are not appreciated very much as "neighbours". Companies which are active in densely populated areas find that complaints by residents via the politics control system are often very restrictive.

The industrial stages represented by materials manufacturers, designers, recycling and disposal companies see themselves confronted with other pressure groups, the interest groups[16] and the scientific institutions. These have an effect primarily via training, information generally and in particular by the publication of research results. As a result of the high degree of specialist authority, these pressure groups have a less spectacular but very long-lasting effect.

2.5 Results of a Prognos study (1999)

In a recent study, Prognos, a research institute, attempted to estimate how 66 of today's well-known energy-saving and resource-preserving products, processes and technologies will affect the employment situation by the year 2020, if the demand for these is so high that the use of non-renewable energy sources is reduced as a result by 30% and the consumption of all other materials by 25%. As a comparison scenario, the study uses a hypothetical situation in 2020, assuming that consumers did not become more conscious about the environment. In addition to the calculated direct employment effects in the various sectors, Prognos also attempted to consider the indirect effects which result from substitution of alternative products and the concomitant effect on incomes and the total demand of households. Table 1 below summarises the most important results of the study.

---

15 According to UVPV art. 2 projects for new installations or substantial conversions of existing installations are subject to environmental compatibility inspection.

16 Examples are: Canton and municipal building departments, environmental specialists which together with the research institutes form a construction/biological network of pan-Swiss significance.

Table 1: Employment effects of environmental innovations in Switzerland

Sector	Direct effects	Indirect effects	Total
Agriculture	0	0	0
Mining	0	0	0
Metal industry	0	0	0
Car construction	-900	-300	-1,200
Electrical engineering	600	100	700
Machine construction	-300	-700	-1,000
Chemicals	500	100	600
Stones and Soil	0	0	0
Construction	**5,000**	**3,000**	**8,000**
(of which thermal insulation)	**3,000**	**1,200**	**4,200**
Electricity, gas and water	500	-200	300
Traffic	5,600	-4,000	1,600
Sub-total	11,000	-2,000	9,000
Total effect			17,900

It is interesting that according to the Prognos study by far the most jobs would be created in the construction industry. If direct and indirect employment effects are combined, almost 90% of the net jobs created are in the construction industry, where in turn more than half can be allocated to thermal insulation. With reference to our case study this is significant in that the companies involved in the production of energy-saving windows are among the most important economic actors in the thermal insulation sector. The Prognos study therefore indicates that the manufacture and installation of energy-saving windows could create a considerable number of jobs in the future.

In contrast with our case study, which analyses development to date, the Prognos study, for its part, is future-oriented and thus based on specific assumptions which are in no way binding. So the estimated employment effects in the individual case must be interpreted with caution. However, the important point remains that for companies which provide services in the field of thermal insulation there does exist a job creation potential which must not be underestimated. Eco-innovations in this area would then also be especially likely to generate positive employment effects.

3 Energy-saving windows as an eco-innovation

In order to assess the impact of eco-innovations on employment and to improve understanding of the mechanisms, the concrete employment effects are evaluated at two window-producing companies.

Eco-innovations which are suitable as objects for investigating their effects on employment must meet two basic criteria:

1. The employment effects must be significant. The subject for investigation must therefore be found in a major branch of the national economy.
2. The case study must be able to take account of the environmental effects and the value chain.

The window as an eco-innovation meets both conditions, as is shown in the sections below. First, the general function and tasks of a window, the possible types of construction and the fundamental value chain are discussed. Secondly, the actual eco-innovation – "energy-saving windows" – is discussed and the decision criterion of energy-saving, or rather thermal conductivity (k-value) is introduced. Finally, in the main part, the possible employment effects of windows as an eco-innovation are indicated, using the two companies EgoKiefer AG and Dörig Fenster AG.

3.1 The value chain

Most windows available on the market today are single windows with multi-pane insulating glass, with frame materials such as wood, plastic, aluminium, aluminium-wood and metal.[17] This section provides a short description of the value chain of these windows.

At the beginning of the value chain we find the extraction, basic processing and transport of the raw materials for the components: frame, glass and seal (in multiple-glazing between the glass and the frame and between the frame and the structure).

In the second stage, the windows are manufactured and transported either directly to the end user or to the market. The production processes vary, depending above all on the frame material. In the case of plastic windows, prefabricated components (profiles) are cut to size and assembled, whilst with wood and wood-aluminium windows the supplied cut timber is brought in various stages into the required shape and then assembled.

---

17 In Appendix I the environmental effects of the individual frame materials are discussed. In this case study, energy-saving during the period of use of a window is used as a measure of the "environmental protection effect".

After manufacture, the windows are packed and transported to site, where they are distributed. Then the new windows are fitted, or the old windows are removed. As a final phase, the recycling of the old windows may be considered; the key point is the recycling options for the individual components.

3.2 Key players in Switzerland

The window manufacturers belong to the association of building finishers. The window manufacturers have a share of 4-5% of the total turnover of the Swiss construction industry which is about 44 billion CHF. The Swiss window market can be subdivided into four groups that possess the following market shares:

- 40% wooden windows, mostly produced by small and very small companies;
- 17% wood-aluminium windows;
- 35% plastic windows;
- 8% metal windows.

The two largest plastic window manufacturers in Switzerland are EgoKiefer AG and Bachmann AG.

In the sector of the wood and wood-aluminium window manufacturers, Fenster-Dörig AG agreed to co-operate, which, in this subsector, is one of the largest companies.

3.3 The eco-innovation of energy-saving windows

The criterion applied to actual eco-innovation is energy-saving through the use of insulating glass windows. Thermal conductivity (the k-value) is available as a measurable criterion for the energy-savings. The lower the k-value the higher is the insulation capacity.

The thermal conductivity of a window (k-value) is calculated from the different values for the following components: structural connection, frame, window leaf, glass and spacers between the layers of glass, and therefore depends very much on the design. Plain wood, wood-aluminium and plastic windows have very similar k-values. Plastic windows meet a standard k-value of 1.3-1.4 with 1.1 k-value glass.

Since a better k-value can be achieved with glass than with the frame, larger windows will have a better thermal conductivity value than small windows under identical conditions.

The most important development in the area of thermal insulation was glass development, namely the introduction of double glazing. A single-glazed window has a thermal conductivity value of around 3. In a first stage, two panes of glass with a sealed gap were used to slightly improve the k-value to approximately 2.6. Only by coating the outer glass panel with an infra-red radiation reflecting layer and filling the gap with a gas (argon) could be realised a substantial improvement in the k-value to approximately 1.1.

Today triple glazing is common, where the two outer panes are coated and krypton is used as the filler gas. This technology means that thermal conductivity values as low as 0.5 can be achieved. However, such glazing is very expensive – above all because of the very expensive krypton gas. These windows are fitted virtually only in so-called low-energy homes. In these houses, the connection of the window with the structure is equally important. At present it is not yet possible to estimate how the standards for "low-energy houses" will be propagated. However, the main drivers or motivators will undoubtedly be legislation on the one hand and energy prices on the other.

3.4 Case study: EgoKiefer AG

Founded in 1932, EgoKiefer today is the largest Swiss window and door manufacturer, with its head office in Altstätten SG. 35% of the entire Swiss window market is for plastic windows and EgoKiefer has a 30% share of this. In 1998, in a production area of 31,000 m^2, 800 employees produced 220,000 windows and 25,000 doors. Sales of over CHF 180 mln were achieved via 9 branch offices, several regional sales offices and over 150 specialist and partner companies.

Turnover has increased continuously since 1986, from CHF 70 mln to CHF 180 mln in 1998.

Concerning the question of whether plastic windows also meet the criterion of an "eco-innovation", the introduction to the case studies indicated that as far as energy-savings are concerned (the k-value), no difference can be established between plastic, wood and wood-aluminium windows. The investigation of the employment effect in the production of plastic windows is therefore justified.

3.4.1 Stages in the production of a window

12 different Polyvinyl chloride (PVC) profiles are supplied in 6 m lengths. By combining the different plastic profiles, EgoKiefer AG is able to construct all forms of rectangular, sloping and arched-shaped windows.

To ensure that the plastic profiles have the required torsional stiffness and jointing at the corners, steel profiles are inserted into the plastic profiles. The steel profiles are also supplied in 6 m lengths.

The 6 metre lengths of PVC profile and the steel profiles are cut to size according to the order. For the last five to ten years all windows have been "made to measure", i.e. fabricated according to the client's wishes. From this point of view, a window is not a standard item.

The steel profiles are inserted into the plastic profiles and the composite profiles thus stiffened are welded to form a window frame or leaf. Depending on the customer's colour preferences, the frames and leaves are then painted[18].

In a subsequent process the fittings – the connections between the window and the frame – are fitted. Finally, the glass is inserted and fixed in the leaf with a clip-on window sealing lip. Thus the window forms a complete system (frame, leaf, glass), which only receives its functionality and torsional stiffness when all its components are assembled.

Once the window has been manufactured it is packed, transported to site, distributed among the various rooms and finally fitted. The graph below represents the production process in summary form for both plastic and wood windows.

3.4.2 Employment effects at EgoKiefer AG

In the last ten years, the number of employees at EgoKiefer has increased from 661 to 800. Of the 800 employees, some 300 work in support functions: administration (finance and personnel), sales, service, and research and development. The other 500 employees work in production and fitting.

As far as the level of training is concerned, virtually all employees in the support functions have completed apprenticeships of a technical or commercial nature. In research and development (5 people), two persons have a technical academic qualification.

In production, 80% are semi-skilled employees and 20% have completed an apprenticeship. In fitting, approximately 50% have a qualification as a carpenter or joiner and 50% are semi-skilled.

18 EgoKiefer finishes its windows with water-soluble paints. However, this does not result in increased sales. Nor are there any other operations or employee qualifications required, from which an employment effect can be derived. The reason for the introduction of water-soluble paints was to conform with environmental requirements (VOC and FCKW).

Figure 4: Production stages of windows

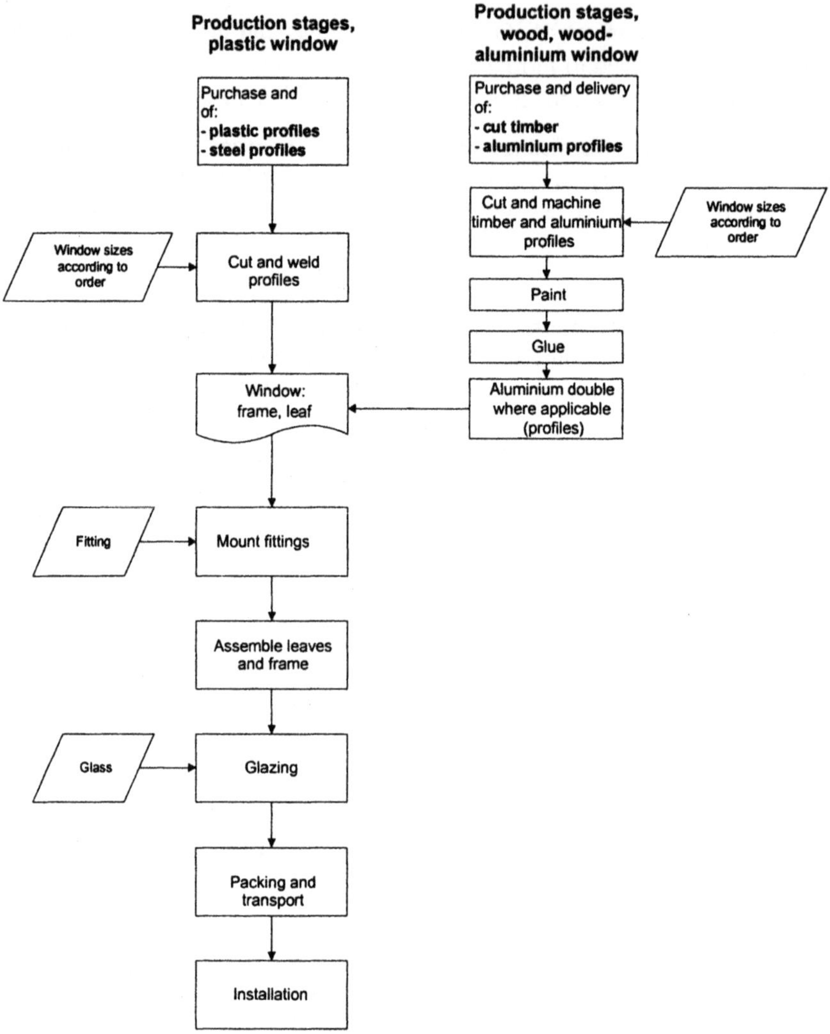

3.4.3 Direct employment effects in the upstream stages

In the production stages upstream, an employment effect could only be identified in glass production.

In 1996, EgoKiefer AG developed Egoverre with a thermoplastic spacer, in conjunction with its German glass supplier Schott Glas AG.

Since then, 80% of EgoKiefer windows have been equipped with Egoverre. For the German glass manufacturer, this made EgoKiefer such an important client that it built a branch factory in Altstätten, creating 40 jobs. However, this positive employment effect is counterbalanced by a reduction in orders for other previous glass suppliers. Therefore, the overall effect in terms of the impact on employment at the glass suppliers is difficult to estimate and is not taken into account below.

The purchased profiles in PVC and steel have changed shape, but the design changes represent a continual development in profile production. One further aspect makes it difficult to estimate the employment effect at the profile manufacturer stage. On the one hand, profile production is a large-scale business, where it is difficult to attribute a number of jobs to specific order quantities. In addition it is hard to estimate how easy it would be for the profile manufacturer to find another client, thereby avoiding any job losses.

3.4.4 Employment effects in downstream stages

The stages following production transport to site, distribution, installation on site and service – are covered on the one hand by EgoKiefer AG, and on the other hand by more than 150 specialist and partner companies.

The employment effects within EgoKiefer AG developed from 192 employees in 1989 to 224 employees in 1998; since 1995 the service and fitting departments have been managed separately.

It is once again impossible to estimate the employment effects in the 150 specialist and partner companies, because of the difficulty in establishing the causal relationship between Egokiefer AG's order book, the general order situation and development trends specific to the industry, in addition to the corresponding effects on the number of employees.

3.4.5 Indirect employment effects

The question of promotion of complementary goods and the substitution of goods regularly presented interviewees with major problems.

In principle one can conclude that as a result of the energy-saving due to energy-saving windows, smaller heating systems can be installed. De facto, however, the window manufacturers maintain that the replacement of old windows with energy-saving windows is not necessarily accompanied by the installation of a smaller heating system. Often existing heating systems are retained while the heat output is reduced.

Complementary goods such as window-sills, sunshades and even doors are for the most part marketed only by large manufacturers. Classification of employees according to the production lines is, however, not possible within EgoKiefer AG, since workers are assigned to different production lines depending on the order situation. Moreover, complementarity relates more to a product offer as a "bundle" or to a complete solution from a single source.

An eco-innovation could also cause indirect employment effects if it affects prices. And, indeed, prices for energy-saving windows have continuously fallen over the past years, which increased the portion of income that could be spent for other goods or services. In this case there would be potential positive indirect employment effects caused by an increase in production activities in other industries. However, the falling price trend was not caused by innovations in the production of windows but rather was the effect of the tough competition on the market for windows. Therefore, income and substitution effects due to falling window prices cannot be related to the production of energy-saving windows.

The question of the employment effects at competitors, finally, is another important question to which this study cannot provide an answer as no corresponding information or statistics are available. Especially important in the case of EgoKiefer AG is the potential future growth of exports to the surrounding countries (mostly Germany) because of the expected changes in construction laws in these countries. In Germany, new construction laws will require lower k-values, which, according to EgoKiefer, cannot be met by many German window producers unless they invest a lot of money in new production technologies. Therefore, the German market is expected to provide additional export opportunities. However, the increase in production of windows at EgoKiefer will probably crowd out German window producers and the net employment effect will again be zero.

3.5 Case study: Fenster Dörig AG

Apart from a small proportion of plastic windows (20%), Fenster Dörig AG produces mainly wood windows (40%) and wood-aluminium windows (40%). With 75 employees, the company has for some time regularly been turning over CHF 14 mln per year.

About 50% of windows are used in new buildings and 50% in renovation projects (70% in detached houses and the remainder in multi-occupancy houses, industrial buildings and offices). The construction market is judged to be stagnating; in this context, the particular regional features of the Appenzell canton – a high proportion of old or historical buildings which have already been renovated, with relatively expensive wooden window structures – must be taken into consideration.

Basically, Dörig Fenster AG pursues a strategy of taking on more small orders than large ones; this is first because of the cluster risk (risk of bad debts), and secondly because of the shortage of capacity or the tying-up of capacity by a large order, and finally because of an attempt to remain independent of customers.

In terms of service provision, Dörig Fenster AG places great value on high-quality windows; only in this way can the 30% difference in price compared with a plastic window be justified to a customer.

The demand for glass with better k-values increased with the introduction of cantonal building regulations. Thanks to higher marketable quantities, this increased demand led to a perceptible price reduction for the glass. Because of this effect, since autumn 1996 Dörig Fenster AG has not been using any glass with a k-value above 1.0. Together with the frame/leaf, this produces a k-value of 1.3-1.4 for a window. The switch from glass with a k-value of 2.9 to glass with a k-value of 1.0 can be described as the actual eco-innovation of Dörig Fenster AG.

The reasons for this were legal requirements (building standards), above all in the Thurgau canton, which has the most stringent standards in Switzerland.

By withdrawing glass with k-values higher than 1.0, Dörig Fenster AG, as far as the building regulations are concerned, is very much on the safe side. In addition, the customer has a window which will undoubtedly meet future standards for years to come. Apart from meeting legal standards, the colour or rather the light transmission of a window also plays a key role. Applying coatings to the glass affects its light transmission which we perceive as a different "colour" or "tint".

This situation persuaded Dörig Fenster AG to convert its production to coated glass with very good insulation properties at an early stage (k-value 1.0). This glass will still be available for some years to come which ensures standardised maintenance for Dörig Fenster AG, and the customer can assume that even if he has to replace the glass, his windows will still be of a uniform "colour".

In financial terms, Dörig Fenster AG are pursuing the goal of keeping sales constant but at the same time – above all in the areas of materials and human resources – reducing costs. However, jobs are not lost in the conventional sense. It is just that with every natural loss of a worker, a consistent check is carried to see which part of the work must be retained, which part of the work can be distributed to the remaining employees, and which part of the previous work can be rationalised. By applying this principle, Fenster Dörig AG have managed to stay afloat in the highly competitive market for wood and wood-aluminium windows.

The actual employment effect at Fenster Dörig AG is therefore a reduction of 10 jobs in ten years.

4 Conclusion: assessment of employment effects

Based on the results of our case study we can try to estimate the employment effects of the adoption of energy-saving windows in the Swiss construction industry. As this eco-innovation may be classified as a product innovation, the hypothesis formulated in the IMPRESS rules on case studies was that we should expect positive direct employment effects, while the indirect employment effects would depend on the degree of substitutability between the conventional product and the new product. Because the substitutability between conventional windows and the new energy-saving window is high, we should probably expect negative indirect employment effects, which can partly or entirely offset the positive direct employment effects by their domination of conventional windows.

Assessing the employment effects requires a benchmark in order to put our employment data from the case study into perspective. The decision made in the IMPRESS rules on case studies is to use the LCA methodology. There, environmental effects of a certain product or service are compared with a functional equivalent unit, meaning, for example, that a product is compared with another one able to deliver a similar function. Therefore, there is a need to identify a functional equivalent for our eco-innovation that will help us to understand the employment effects. In the IMPRESS rules on case studies the following definitions of functional equivalence are considered to be applicable for our purposes:

1. The situation before the innovation was introduced.
2. A similar context where the innovation did not take place.

We will use both definitions of a functional equivalence in our assessment of the employment effects. However, we have to define the "similar context where the innovation did not take place" more precisely. As energy-saving windows are the only windows available on the market now, we cannot compare our innovating firms with "non-innovating firms" in order to find out whether there is any significant difference in the development of employment. Consequently, we have to choose a different approach. As the development of construction activities as well as of employment in the total construction industry is well documented in Switzerland, we will try to assess the employment effect by relating the development of employment in the innovating firms to the development of the whole construction industry.

Looking first at the development of employment in the innovating firm and, therefore, using the situation before the innovation was introduced as our reference case, we can clearly establish a positive direct employment effect in our first example (EgoKiefer AG). The main innovation activities that led to the production of the new energy-saving windows took place between 1986 and 1989 and since then, the company has produced only these types of windows. In the ten years that followed (1989-1999) total employment increased by 19% from 673 in 1989 to 800 in 1999. A similar increase can be observed if we concentrate on the employees that are engaged in the distribution/fitting of the windows as well as in service activities. The number of these employees increased from 192 in 1989 to 224 in 1999, which corresponds to an increase of 17%. Significant direct employment effects upstream and downstream in the value chain could not be detected and they do not seem to be of relevance.

In our second example (Fenster Dörig AG) no positive employment effects can be observed. However, the major innovating activity (production of energy-saving windows with k-values as low as 1) in this firm occurred only three years ago and, as our first example shows, employment effects may not be felt immediately after the innovation. Firms will not hire additional people unless they are quite sure that they can sell more of their products over a longer time span. Furthermore, construction activities are highly cyclical and the overall economic situation (the business cycle) can exert a large influence on the decision whether to hire additional workers or not. Consequently, we will concentrate the assessment of the employment effects on our first example (EgoKiefer AG) as no significant insight can be gained from our second example. But we have to keep in mind that the majority of the firms dealing with the production and installation of windows are locally-based small and medium seized enterprises and the example of EgoKiefer AG, which is one of the key players in the window market, may not be representative of many smaller companies.

Furthermore, we put the observed direct employment effects in our first example (EgoKiefer AG) into perspective by identifying a similar context where the innovation did not take place. In this context, the relevant question is: which part of the observed increase in employment can actually be attributed to the eco-innovation and which part of the increase would also have occurred without the innovation? In order to give a reasonable answer we cannot avoid making some assumptions based on observable trends in the development of construction activity in Switzerland. Total construction activity in Switzerland reached its peak in 1990 and has since shown a declining trend. This negative trend since 1990 is also reflected in the development of employment in the sector that declined by 11.3% between 1991 and 1995 (the last year for which data is available). Therefore, taking these figures into account the employment effect at EgoKiefer appears even more impressive and interpretation of the increase in jobs between 1989 and 1999 probably understates the actual impact on employment.

However, a further aspect has to be taken into consideration. As current Swiss laws do not permit the installation of any windows other than energy-saving windows in new buildings, the share of energy-saving windows that are installed in new buildings cannot be considered to have any positive net employment effect in this case. The positive direct employment effect is fully compensated by a negative indirect employment effect because no additional demand is induced by the eco-innovation. Instead of buying non-energy-saving windows, as was previously the case, customers now buy energy-saving windows. But, as became evident from our case studies, the production of energy-saving windows is not more labour-intensive than the production of other windows and therefore there is no positive net employment effect.

Positive net employment effects can only be attributed to the energy-saving windows that are installed in order to replace conventional windows in existing buildings. In this case, we have an additional demand induced by the eco-innovation because, otherwise, in most cases there would not have been any incentive to replace the conventional windows. Unfortunately, EgoKiefer has no data that would allow us to break down the demand into the demand for windows to be installed in new buildings and the demand for windows to be installed in existing buildings. However, it seems reasonable to assume that the percentage of windows demanded for installation in existing buildings is similar to the percentage of expenditure for renovation of existing buildings of total construction expenditure in the Swiss economy. Between 1989 and 1999 the average value was 35%, meaning that 35% of the money spent on construction was spent on renovation activities and 65% was spent on the construction of new buildings. Based on these figures our best estimate is that 35% of the 127 jobs that were created between 1989 and 1999 can be attributed to energy-saving windows. This estimate may still be overly optimistic because some windows that were replaced would actually also have been replaced by new conventional windows if they were still available on the market. But even then we can establish a positive net employment effect in our example. At least in the case of EgoKiefer AG, the eco-innovation under investigation also had a positive impact on employment.

Finally, it should be mentioned that the introduction of energy-saving windows in Switzerland was clearly driven by politics. The main reason why house owners choose to install energy-saving windows is simply because they have to. The relevant laws on construction have gradually been adapted in a way that makes the installation of non-energy-saving windows almost impossible nowadays. Swiss construction laws are probably some of the strictest in the world and have a strong influence on decision making regarding the choice of windows and other insulation devices when building or renovating houses. The market, on the other hand, did not provide any motivation to install energy-saving windows. As energy prices have been at a comparatively low level since the late 1980s, energy-saving is not a major goal of most households and companies.

References

Baccini, P, H. Belevi and T. Lichtensteiger (1991), *Die Deponie in einer ökologisch orientierten Volkswirtschaft. Eidg. Anstalt für Wasserversorgung, Abwasserreinigung und Gewässerschutz (EAWAG)*, FER-Studie für den Schweizer Wissenschaftsrat, Dübendorf.

Bundesamt für Statistik (1986), *Erhebungen über die Gütertransporte auf der Strasse 1984*, Bern.

Dyllick, Th. (1992), Ökologisch bewusste Unternehmensführung. Bausteine einer Konzeption. *Die Unternehmung* Vol. 46, No. 6, 391-413.

Dyllick, Th. (1994), *Ökologischer Wandel in Schweizer Branchen*, Bern.

Gesundheits- und Wirtschaftsamt der Stadt Zürich (Ed.) (1992), *Umweltbericht 1990/91, Stadt Zürich*, Umweltschutzfachstelle, Gesundheits- und Wirtschaftsamt der Stadt Zürich, Zürich.

OECD (1998), *Umweltschutz und Beschäftigung*, Paris.

Hugenschmidt, H. (1993), *Ökologischer Branchenstrukturwandel in der Schweizer Güterverkehrsbranche,* Diskussionspapier, Institut für Wirtschaft und Ökologie Hochschule St. Gallen (IWÖ). St. Gallen.

IP Bau, Bundesamt für Konjunkturfragen (Eds.) (1992), *Umweltbewusst Bauen. Leitfaden für umweltgerechte Erneuerung,* Bern.

Keller, B. (1992), Energiebedarfsbestimmende Eigenschaften, in: ETH Zürich, Abteilung für Architektur, *Ökologie am Bau,* Zürich, 41-65.

Koller, F. (1994), Ökologischer Strukturwandel in der Schweizer (Hoch-)Baubranche, in: Dyllick, Th. (Ed.), *Ökologischer Wandel in Schweizer Branchen*, Bern.

Prognos (1999), *Mehr Arbeitsplätze durch ökologisches Wirtschaften? – Eine Untersuchung für Deutschland, die Schweiz und Österreich*, Köln.

Schwank, O., P. Koch and U. Mauch (1992), *Struktureffekte durch umweltverträgliche Abfallbewirtschaftung. Fallstudien in ausgewählten Branchen. Strukturberichterstattung,* Studienreihen vom BFK, Bern.

Wüest, H. and Ch. Gabathuler (1990), *Bauland Schweiz. Grundlagen und Perspektiven zum Bau- und Baulandmarkt und zur Siedlungsentwicklung der 90er Jahre*, Zürich.

Appendix: ecological assessment of frame materials

Frame material: wood

Domestic wood is the traditional framex material for windows in Switzerland. Other window materials became significant only after the Second World War following the growth in oil and aluminium processing industries. Tropical timbers have been extensively used as window materials only since the mid-1960s. A distinction is made between tropical and domestic wood, as well as the condition of the raw material, suitability for use and economy.

In practice, the following types of timber are most used today: Spruce, pine, larch, Oregon pine, hemlock, types of red meranti and mahogany.

Stages in the production of a wood window frame

1. Cultivation
2. Felling
3. Transport to sawmill
4. Sawing, drying, profiling, assembly, coating

Employment effects

Skilled labour in various areas of carpentry, from the sawmill via the joinery to the window fitter.

Recycling

An intact window can be re-used.

Otherwise separation of the other components and processing of frame into shavings or chips (chipboard, wood fuel).

Frame material: plastic

90% of plastic windows are made from polyvinyl chloride (PVC). Raw materials for production are crude oil and chlorine chemistry products.

The process of oil extraction and chlorine production involves great ecological risks.

Further processing of PVC to make window profiles requires the use of some problematic ancillary materials (e.g. stabilisers containing lead).

Recycling

Currently, recycling of plastic windows is not possible, only downcycling to cheaper products e.g. flower boxes.

Frame material: aluminium

The bauxite used for aluminium extraction is mined mainly in Australia, Guinea, Jamaica and Brazil and is associated with major depredation of the landscape where it is processed into aluminium oxide[19].

To obtain 1 ton of smelting aluminium, one needs:

- 4t bauxite
- 0.5t electrodes
- 0.05t cryolite
- approx. 72,500 kWh of energy

Cryolite electrolysis releases highly toxic fluorine, which is very poisonous for cells and plants.

Recycling

Recycling of aluminium window frames is technically feasible. The energy expenditure is reduced to 5-10% of the original requirement. However, substantial pollutants are also produced during recycling, particularly due to coating of the window profile with acrylic paints. The table below summarises the ecological and economic assessment of window frame materials once more in abbreviated form:[20]

19 Producing one ton of aluminium oxide creates 2 tons of red mud containing iron oxide which must be disposed of in a tip.

20 Based on LBB, 1993: 29.

Table 2: Ecological assessment of different window frame materials

Assessment criteria	Wood					PVC	Alu	Alu-wood
	Home-grown			Tropical				
	Open-pore coating	Surface coating	Alkaline treated	Open-pore coating	Surface coating			
Obtaining raw materials								
Environmentally sustainable	0			-		-	-	0
Manufacture								
Environmentally sustainable	+			+		--	--	-
Energy consumption	+			0		-	--	-
Suitability for purpose								
- Durability	+	+	+	++	++	0	++	++
- Function	+	+	+	+	+	+	+	+
- Upkeep	0	+	0	0	+	+	+	+
- Repair	++	++	++	++	++	-	0	+
Re-use								
Recycling possibility	-	-	-	-	-	-	++	+
Downcycling possibility	+	+	+	+	+	+	++	+
Environmental sustainability	0	-	+	0	-	--	--	-
Disposal								
Incineration / dump	0	0	+	0	0	--	0	0

Key: ++ very positive, + positive, 0 neutral, - negative, -- very negative

Recycling

Suhita Osório-Peters

1 General aspects

The topical questions related to recycling have recently been formulated by the EU-Commission: "Recycling is not only an environmental priority but is also intrinsically profitable in an increasing number of applications, thanks to energy-savings and economies in materials and of other types which recycling permits compared with traditional processes. It is vital, therefore, that the framework conditions (...) do not hinder the development of these activities and, on the contrary, that Community action takes due account of the economic and job creation potential of recycling activities" (European Commission, 1998: 1). However, the EU-Commission recognises that recycling, especially the recycling of steel scrap and non-ferrous metals, has existed for a long time and experience shows that "in all cases, the growth of the traditional recycling industries has encountered limitations sooner rather than later, depending on the different materials concerned, in the form of insufficient demand, precarious and, for the most part, not very competitive market structures and of unbalanced competition from virgin raw materials" (ibidem). Increasing awareness of the environment, political regulation of waste management, and growing demand on the part of society for recycled products have, however, helped to boost a renaissance of recycling not only as an instrument to attain long-term environmental objectives but also as an "example of the complementary character of industrial development and environmental protection" (ibidem).

For the purposes of this study we will use the definition of the European directive on packaging and packaging waste: "recycling" is the *reprocessing in a production process of the waste materials* for the original purpose or for other purposes, including organic recycling but *excluding energy recovery*. "Recycling industries" also include companies with activities such as collecting, dismantling and sorting provided they are covered by the definition of recycling in the sense of the directive mentioned above (European Commission, 1998: III).

Though recycling is normally expected to create jobs, the simple re-use of material residues within a production process (closing material cycles with integrated technologies) is not seen by experts as a relevant job creation source. The scant information available about some real life examples seems to bear this out. In contrast, recycling activities consisting of additional stages in the value chain of waste substances or end-of-life products are theoretically associated with positive employment effects.

As part of the IMPRESS project, this case study aims to achieve two main goals: first, it is meant to clarify the motivations and decisions of individual eco-innovative firms and their impacts on employment. Second, it is designed to be used as an explorative analysis to test the adequacy and limits of the proposed methodology (value chain method) for the purposes of the study. Moreover, though the results of a case study cannot simply be extrapolated, the case study is intended to produce some empirical basis for a better understanding of the relationship between employment and recycling in general.

In the study at hand, only one part of the wide-ranging recycling sector will be studied in detail: the recycling of CFC-refrigerators, which (from an environmental point of view) falls into the electronic scrap recycling industry. But, for a better understanding of the economic and political framework of this specific case, the industry and employment structures of the whole recycling sector in Germany will be outlined as will the most important political issues, especially those concerning the recycling of electronic scrap.

2 The recycling sector in Germany

2.1 Sector structure and literature review

It is extremely difficult to provide a complete description of the recycling industry, because the boundaries between production and recycling and between recycling and other waste management activities are not clearly defined. Accordingly, it is not possible to get reliable information about employment in this sector. Empirical studies on the impact of environmental protection on employment are normally based on estimations of the labour allocated to environmental protection in private firms or public institutions in relation to the costs of or investments in environmental protection (reduction of emissions), but they do not analyse the recycling activities separately (cf. Umweltbundesamt, 1993: 14-17, Umweltbundesamt, 1997: 50). According to one of the rare studies devoted to the recycling industry, the number of employees in private firms dealing with the collection, sorting and treatment of waste materials in Germany may be estimated at about 162,000 in 1992/93 (RWI, 1999: 19). Apart from this, most of the sketchy information on the number, size, capacities and employment figures of firms that state they deal with recycling activities is based on surveys in the industrial and service sectors (MUP – Mannheimer Unternehmenspanel, Stiftung Arbeit und Umwelt, 1999, BVSE, 1999[1]).

Recently (in October 1999), on behalf of Duales System Deutschland (DSD) – the most important institution dealing with the recycling of domestic waste materials in Germany – and the Stiftung Arbeit und Umwelt (an institute of the trade union IG Bergbau, Chemie, Energie) carried out a study on the employment effects of the "Commercial and Industrial Waste Management Act" (the Kreislaufwirtschaftsgesetz), which was passed in 1994 and came into force in Germany in 1996. It provides for a reduction of the material throughput in the economy by closing material cycles. The law focuses especially on the fields of recycling that are already regulated by law (packaging, batteries, organic waste, paper, glass) and those that are expected to be regulated in the near future (end-of-life vehicles, electrical and electronic waste). The study states the current employment figures in the recycling industry, and also the future development of the "Closed Cycle Economy" in Germany, assuming that the legal principle of "producer responsibility" is completely implemented. The results of this study are summarized in Table 1.

Table 1: Employment in material cycles

Materials Status 1997/98	Potential assuming complete producer responsibility
Light packaging* 16,650	already realised
Paper packaging* 1,800	already realised
Glass packaging* 1,850	already realised
Batteries 550	already realised
Cars 8,100	6,900
Electronic scrap 6,500	2,500
Graphic papers 4,900	no changes expected
Organic waste** 3,000-4,000	producer responsibility not possible
Total ca. 44,500	ca. 54,000

* including jobs in the DSD; ** excluding collection , Source: Stiftung Arbeit und Umwelt (1999: 16).

1 Both the study carried out by the Stiftung Arbeit und Umwelt and the data published by the Bundesverband Sekundärrohstoffe und Entsorgung e.V. (BVSE) are intended to support specific interests in the recycling sector, thus some caution is called for. The source of the MUP data is Credit Reform, a politically independent institution that collects raw data from official registers and direct surveys.

On the basis of the above mentioned ZEW database MUP, which records all the firms that have been founded in Germany from 1990 on, we tried to draw up the current profile for firms that were listed under the heading "Recycling" (in September 1999). For this calculation we took the data of 1,366 newcomers (1,235 firms in Western and 931 in Eastern Germany) employing 11,682 and 11,407 persons, respectively (Table 2).

Table 2: Number of firms in recycling in Germany (1999)

	West Germany	**East Germany**
scrap recycling		
ferrous metals	251	191
non-ferrous metals	49	47
Non-metal recycling		
recycling of plastic waste	272	116
recycling of other waste materials (excluding textiles, paper and glass)	663	577
Recycling (total)	**1,235**	**931**

Source: MUP – ZEW.

Most employees in this business sector can be found in non-ferrous and ferrous metal, paper and plastic waste recycling. Scrap, non-metal material and textiles recycling enterprises have clearly less employees.

In Western and in Eastern Germany, the recycling sector is dominated by a large number of small firms (80% of the firms in Western Germany and 70% in the Eastern "Länder" have fewer than 20 employees). In Eastern Germany, relatively speaking, there are more medium-sized and large firms (with 50-500 employees) than in the West (Figure 1). This is striking, because on average the enterprises in East Germany are significantly smaller than those in West Germany.

Figure 1: Firm size in the recycling industry

Source: MUP – ZEW, 1999.

Though all our interview partners saw a strong trend towards concentration in the recycling sector, these findings suggest that the traditional structure of the industry has not changed significantly during the last 10 years. This apparent contradiction is probably due to the fact that a lot of intermediary waste traders have penetrated the sector during the last few years and most of our interview partners regard recycling as an industrial activity (i.e. as the transformation of materials); in this field of activity concentration is probably really increasing. A large number of small firms have been taken over by stronger competitors.

2.2 Recycling of electronic and electrical scrap in Germany

Estimations of the volume of electrical and electronic scrap (EES) can be based on two different methods (cf. Halstrick-Schwenk, 1999: 86-94):

a) on the so-called *life-cycle method* that takes into account the number of electrical/electronic products sold at a given moment as well as the assumed average life of these products

b) on the *end-of-life method* that takes the actual average share of EES in waste as a point of departure for extrapolation.

Legislation in Germany and the EU has normally been based on the first method, which was also used in the so-called Töpfer Study on EES, which was roughly confirmed by estimations made by the Zentralverband Elektrotechnik und Elektronikindustrie (ZVEI) (a producers' trade association). Using the life-cycle method they put the annual volume of electronic scrap in Germany at 1.5 mln tons (ZVEI) and 1.8 mln tons (Töpfer), respectively (Hafkesbrink, 1999: 171).

Estimations based on the second method were made by the Federal Statistical Office (Statistisches Bundesamt), which arrived at a considerably lower amount of EES: 210,000 tons. This divergence was confirmed in 1996 by Interseroh AG, which had conducted a comprehensive empirical study and came to the conclusion that in German households there is a potential volume of 13.5 kg of electronic end-of-life products per person and year. However, only 5 kg of electronic scrap for recycling per person and year (i.e. a total of 310,000 tons) is disposed of as EES. The study showed that one reason for the discrepancy between the prognosis and reality is the fact that consumers use their equipment longer than expected; moreover, with the emergence of an active second-hand market for used electrical equipment and increasing exports to less developed countries, there is today much less EES in Germany[2] than expected. Finally, a written company survey carried out by the Rheinisch-Westfälisches Institut für Wirtschaftsforschung (RWI) at the end of 1996 came to the conclusion that every year a maximum of 460,000 tons of electronic scrap is recycled in about 300-400 firms with 4,600 employees (Hafkesbrink, 1999: 172). We will return to this point later, when we discuss the political issues related to electronic scrap.

2.3 Political issues

Almost 10 years ago, the German government announced an ordinance on electronic scrap which was meant to stipulate that producers are responsible for recovering and recycling the end-of-life products they produce. Against this background, the IT-trade association published voluntary self-commitments specifying the degree and way its members intended to achieve the goals of the ordinance (Halstrick-Schwenk, 1999: 101). However, ten years later, it is evident that they have not managed to set up a comprehensive, functioning recycling system for electronic scrap in Germany. A first draft for a government ordinance was published in 1991. The government's proposal failed, because it lacked a financing model for the recovery of end-of-life products. Finally, an ordinance was passed by parliament in 1998 that only covered IT-scrap (information and communications technology, which accounts for about 7% - 10% of all the electronic scrap, Halstrick-Schwenk, 1999: 82).

However, the announcement of the government's willingness to intervene if all else fails, boosted private initiative and a market for the recovery, recycling and

---

2 Obviously, consumers prefer a longer product life than forecasted. Because of this development fewer jobs were created in the waste management and recycling sectors but it has probably triggered the creation of new jobs, especially in other service industries, such as repairing and trade.

re-use of electronic end-of-life products arose, which has mainly been organised on the basis of regional networks. Based on the official prognosis on the development of the volume of electronic scrap (see above), industry reacted immediately and created recycling capacities, as Figure 2 shows. In 1993, the members of the Bundesverband der Deutschen Entsorgungswirtschaft e.V. (BDE) alone accounted for recycling capacities of 270,000 tons. In the same year, additional capacities amounting to 230,000 tons were being planned or under construction. But in 1996 only 460,000 tons of electronic scrap were actually treated in Germany.

Public-private regional networks for waste management have been established in almost all “Länder”. Their respective institutional arrangements not only regulate the way the recycling process is organised and the allocation of tasks and responsibilities to private and public actors. Often they are also used as political instruments for local employment policies, because the recycling sector offers numerous jobs for a particularly difficult group of unemployed people: handicapped, poorly-skilled and long-term unemployed persons. At least partially subsidised by the government, these jobs have become an important cost advantage for small recycling firms. Several interview partners even expressed the opinion that public employment subsidies have been one of the main factors creating successful, environment-friendly recycling in Germany.

Figure 2: Recycling capacities in Germany

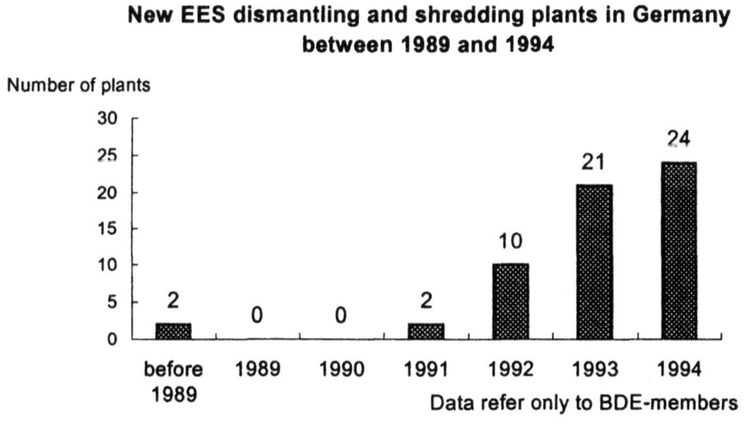

Source: Stiftung Arbeit und Umwelt, 1999: 54.

In fact, in particular with regard to the recycling of end-of-life refrigerators, institutional arrangements between municipalities (that have up to now been legally responsible for recovering end-of-life equipment from households and have had to ensure that environmental recycling standards are met) and private operators have attained quite a high level of CFC-recovery. At first, according to the legal provisions on the ban on CFCs, municipalities were obliged to look for firms complying with fairly high ecological standards. During the first half of the 1990s, the public awareness of the ozone layer problem and quality-oriented competition in the market of environmental services favoured the establishment and diffusion of high technical standards in the refrigerator recycling sector. Later, due to mounting pressure in the market, companies increasingly opted for labour-saving, less expensive (albeit environmentally questionable) alternatives. Currently, environmentally friendly recycling of CFC-refrigerators is experiencing a crisis: though in many cases great efforts have been made to continue to ensure high ecological standards, there is no doubt that some environmentally efficient technologies have suffered as a result of strong market pressure exerted by cheaper disposal alternatives from competitors from Germany and abroad, especially from Eastern Europe.

Currently the discussion is focusing on the extent to which and on how "producer responsibility" is supposed to be legally implemented. Today's situation is totally different from the situation a decade ago, when the first drafts for an ordinance were drawn up. Now, the producers are calling for a system where the end-of-life product returns directly to the producer, or for the establishment of a centralised system similar to the Duales System for packaging waste in Germany, because they have had time to adapt to this form of regulation and now stand to gain some of the profits resulting from the positive development of the recycling industry in the future. In contrast, hundreds of small and medium-sized enterprises (SME) in the recycling industry are calling for a system that takes into account the regionally organised networks and the employment effects that will result from a change in the structure of the system.

2.4 Employment

With regard to employment in the recycling industry, the trade association Bundesverband Sekundärrohstoffe und Entsorgung, e.V. (BVSE) maintains that this sector currently accounts for at least 10,000 jobs in Germany. According to the BVSE, this estimation is based on an exhaustive survey of the association's members that was carried out in June 1999 and had a response rate of 80%. The Stiftung für Arbeit und Umwelt, as mentioned above, estimates that for scrap recycling alone 6,500 employees are allocated to recycling activities in Germany (Stiftung für Arbeit und Umwelt, 1999: 52). As these studies are mainly based on internal data and direct surveys, the differences in their results may be attributable to different calculation bases, but the differences also show the uncertainty surrounding quantitative estimations in this field. On the basis of a company survey the RWI estimated that in 1996 there were about 4,600 jobs in electronic scrap recycling. This means that it represents 5% of the total employment in the recycling industry and only 0.013 of the total employment in Germany. If these estimations are correct, the direct employment effects of the separate collection and treatment of electronic scrap may be considered insignificant for the economy; in contrast, the indirect effects may be important, especially with regard to the production of capital goods for this industry during the period 1991-1995 (Halstrick-Schwenk and Löbbe, 1999: 617). No matter what statistical method is used, the amount of electronic scrap is expected to increase in the European Union and in Germany (BVSE, 1998: 6). In Germany, the estimated total amount of electronic scrap (1.8 mln t/a) is split into the following groups (Table 3):

Table 3: EES in Germany (in 1,000 t/a)

Large electrodomestics (including refrigerators)	630
Home entertainment products	400
EDP equipment	110
Electronic office equipment	110
Communications equipment	140
Industrial electronic equipment	360
Medical electronic equipment	50

Source: BVSE, 1998: 14.

An important part of the recycling activities relating to large electrodomestics – separate collection and transport – is mostly ensured by public institutions (in 412 of the 434 waste management regions in Germany, municipal or regional public authorities are charged with the recovery of this type of equipment). In contrast, small electronic devices from households are often thrown away together with domestic waste and are sent to landfills or waste incinerators for disposal despite their risk potential.

For the purpose of economic analysis, we can distinguish between electronic/electrical waste which retains an intrinsic economic value that can be profitably exploited through recycling and electronic/electrical waste which cannot finance its own recovery and recycling (IPTS, 1996: 39). In this context it is worth mentioning that currently the disposal and recycling costs of electronic waste materials are (at least partially) covered by the waste treatment charges that private households (as end consumers) have to pay to municipalities (for one refrigerator the costs are about € 15-20). In the future, legislation will probably put more emphasis on "producer responsibility". Furthermore, a large number of publicly supported, non-profit organisations are carrying out waste management and recycling activities: In fact, about 1,200 handicapped persons are currently working in 150 workshops dismantling and sorting electronic scrap; another 2,200 jobs are supported by public institutions (Stiftung für Umwelt, 1999: 52).

3 Refrigerator recycling in germany: general aspects

The situation in the sub-sector "Refrigerators" is somehow different from that of EES. Because of the globally successful banning of CFCs through compulsory legal measures, the production of and trade in CFCs stopped in 1995, but old equipment containing CFCs is still in use in Western Germany[3] (see below). The Federal Environment Agency (UBA) estimates that a refrigerator normally has a lifetime of 15-20 years, thus the agency forecasts that at least until 2010 significant quantities of CFC-cooling equipment will be recycled, especially in Western Germany (several studies estimated that between 2.3 and 3 mln end-of-life CFC-cooling products would be recycled in 1993, IFEU, 1997: 7). Estimations based on public statistics stated that 2.2 mln end-of-life refrigerators were collected by German municipalities in 1993. Of these 260,000 were then delivered to firms dealing with waste materials, 1.8 mln to recycling firms, 35,000 to shredding plants, 4,800 were taken to other countries and 23,000 were disposed of in landfills (Halstrick-Schwenk, 1999: 91). On the basis of these estimations (IFEU, 1997: 7), significant capacities for refrigerator recycling have been created in Germany in the 1990s. Because of strong competition in the sector, it is extremely difficult to get objective information from firms. However, it seems that the prognosis has not completely come true. Though currently the amount of end-of-life refrigerators collected by municipalities is likely to correspond to the prognosis (i.e., over 2 mln a year), some of the experts interviewed assume that probably no more than 600,000 refrigerators are actually recycled. Pessimistic forecasters even believe that only about 350,000 refrigerators remain in Germany for recycling, the rest is sent abroad or is disposed of in normal waste incinerators. Though they only used subjective estimations, some interview partners think that only 100,000 of them are recycled according to the highest ecological standards. While there was no consensus among the interview partners about the market volume of this sector, they all put the current number of "real" refrigerator recycling firms at about 30.

The transport, dismantling and treatment of end-of-life refrigerators require special care in order to prevent CFCs from escaping from cooling aggregates and insulating foams into the atmosphere. Thus, besides the requirements for the recycling of electronic scrap in general, for environmental reasons refrigerator recy-

---

3 In the new "Länder", only small amounts of equipment containing CFCs as a cooling agent have to be recycled as no CFCs were used in the production of insulating foams in the former GDR. Currently, only about 3% of the end-of-life refrigerators in Western Germany have cooling aggregates on the basis of ammonia.

cling has to be subject to specific technical standards governing the amount of CFCs that is actually recovered and disposed of. As the law provides for a complete recovery and destruction of CFCs, no legal thresholds for a minimum CFC recovery have been established. However, public institutions like the national environment agency (UBA) and Deutsches Institut für Gütesicherung und Kennzeichnung (RAL) have set some technical standards. The RAL quality criteria are more demanding than the UBA[4] ones, for the RAL-evaluation takes not only the standards in the recycling firm itself into account, but also those in the whole recycling chain. Furthermore, the RAL-labelling criteria are expected to be dynamically adapted to future technical improvements. Firms observing those standards can get a RAL-label. The UBA and/or RAL-standards are also regarded as a guideline for recycling firms participating in the EU eco-audit system. Meanwhile, important trade associations (VDMA, ZVEI and BDE) have also demanded that their members set their ecological standards according to DIN ISO 9000ff.

In the following parts, a specific case study will be examined in detail. Against the background of the political issues and the recent history of the EES recycling, WHITE Ltd.[5], the object of this case study, was selected, because it is a typical example of eco-innovators in this industry. Thus, an extrapolation of the results of this analysis to a larger group of firms in the recycling industry seems permissible.

Before we focus on this specific case, some remarks about the methodology of the case study will be made in the next part.

4 Aims of the IMPRESS case study on recycling

The present case study deals with the recycling of end-of-life products (refrigerators) using a technology that allows different waste materials to be recovered for recycling (metals, plastics, glass), including hazardous substances such as chlorofluorocarbons (CFCs), which are later split into their original chemical components in a special plant.

Recycling itself (closing material cycles) is only an innovation from the perspective of the enterprise which introduces recycling activities. For firms involved in recycling, an eco-innovation has to be defined as new institutional arrangements, or changes in processes and/or products with greater environmental efficiency (see below). This case study will focus on a specific innovative activity in the recycling

---

4 Experts estimate that the UBA standards from 1998 correspond to 90% of the "state of the technique" in Germany.

5 To keep the data and identity of the interview partners confidential, the names of firms were changed.

industry – two-stage recycling of CFC-refrigerators in Germany – which is considered beneficial to the environment.

The hypothesis that recycling is a job-creating activity will be analysed in this context under particular restrictions, since our attention will not focus on the relationship between recycling and employment as a whole, but on the impacts of a specific eco-innovation on the quality/quantity of employment in the respective value chain (possibly ranging from the acquisition of raw materials to the final disposal). Against this background, several questions will be addressed:

a) are the environmentally positive impacts of eco-innovations in recycling coupled with positive employment effects?

b) is it possible to distinguish between direct and indirect employment effects using the methodology proposed (value chain approach)?

c) how do employment effects of an eco-innovation change over the years?

d) would the environmental/employment impacts exist if there had not been any political intervention in favour of recycling?

5 Method and structure of the case study

As was said earlier, this case study is meant to be used as an explorative analysis to test the adequacy and limits of the methodology applied for the purposes of this study. First, the analysis will focus as far as possible on the quantitative assessment of the net employment effects, including the qualitative differentiation of employment (skilled, unskilled manpower). Later the results of this assessment will be discussed in order to categorise them according to the general methodology of the IMPRESS research project (i.e. direct and indirect effects).

Using a value chain approach, the following steps have been addressed:

- selecting an eco-innovation in the recycling industry. Identifying related innovations;

- describing the firm that is at the centre of the analysis. The state the business was in when the innovation was introduced and the subsequent business development; the observation period should be long enough to show the dynamic process resulting from an innovation;

- determining the employment effects in the innovating company during this period;

- drawing up the value chain that emerges from a specific innovation, giving special attention to the system boundaries and describing every point of the value transformation. For this, a methodology similar to a life-cycle-assessment (LCA)-analysis was used;

- fitting the empirical information collected into the context of the general and sector-specific development: motivation of innovation, relationship between innovation and employment and discussion of the results.

Direct interviews with firms were the main source of information for this case study. One research institute (Ifeu, Heidelberg) and eight firms (see list of interview partners in the annex) were interviewed. The most important interview partner was the managing director of the department "Refrigerator-/PUR-Recycling" at WHITE Ltd.

6 Selecting an eco-innovation

The decision on the main innovation for the two-stage recycling of CFC-refrigerators in Germany was taken on the basis of the following selection criteria:

- it is an eco-innovation (see below),
- employment effects are expected,
- information is available.

IMPRESS case studies focus on technological and organisational *eco-innovations*. However, so far there is no unanimity about the definition of eco-innovation. For this specific case study *innovation* is defined from the perspective of the micro-economic actors (enterprises), thus, a change in processes, product or organisation is an *innovation* when it is *new* in the enterprise or sector observed. Furthermore, we defined the ecological quality of an innovation with the following criterion: a process or product innovation is an *eco-innovation*, when it has been positively evaluated by a LCA methodologically based on ISO/DIS 14040. The motivation of the enterprise for introducing this innovation is, from this point of view, not essential.

6.1 Digression: LCA of refrigerator recycling

As part of the "LCA of Waste Disposal", a comparative study on the ecological aspects of different refrigerator recycling techniques was commissioned by the UBA (Federal Environment Agency). The Ifeu-Institut in Heidelberg carried out this study in 1997. The eco-assessment shows that the recycling of a refrigerator consists of several main paths related to the recycling of the following materials (in parentheses the average amount for one refrigerator):

- ferrous metals (16. 8 kg),

- non-ferrous metals (2.7 kg),
- mixed plastics (6 kg),
- insulating materials (CFC-R 11. 300 g) and
- cooling agent CFC-R 12, R 22 etc. (120 g).

One of the most important criteria for the eco-assessment of the disposal of CFC-equipment is the quantity of CFCs that is recovered and neutralised. In the Ifeu study, only two techniques for the neutralisation of CFCs were considered ecologically acceptable: High temperature incineration in special plants with energy recovery (Siemens-Method) and/or the splitting of CFC molecules into their original chemical components, HF, HCl and CO_2 (Solvay-Method). However, only the chemical or thermal elimination of CFC-R 12 has been assessed.

Though insulating foams consisting of CFC-R 11 can be recycled for re-use, they are normally disposed of by incinerating them together with domestic waste. But this option is risky, because CFC is heat-resistant below 1,000°C, as analyses of gaseous emissions from waste incineration plants have demonstrated.

Assuming that the recovered CFCs were properly disposed of and/or treated for re-use, the Ifeu study assessed six different cases:

1. Manual dismantling and sorting of reusable parts
2. Manual dismantling, sorting and re-use of different materials
3. Mechanical dismantling, sorting and specific re-use of different materials
4. Shredding, recycling of the metal parts in the steel industry, incineration of the plastic parts with energy recovery
5. Shredding and incineration in blast furnace
6. Thermal treatment in a "Schwel-Brenn" (literal translation: smoulder-burn) plant (Siemens-Method)

Conclusions from the comparative LCA carried out by the Ifeu-Institut (1997):

- The labour-intensive techniques 1 and 2 were given negative assessments in the LCA-study, because significant quantities of CFC-R 12 were lost during the manual dismantling. Technique 1 is primarily used in Eastern Europe.
- Techniques 4 and 5 were also rejected for technical reasons partially to do with the recycling of the mixed metal components.
- Technique 3 was considered ecologically acceptable, if the recovered CFCs were properly treated.
- Technique 6 (Siemens-Method) was evaluated in a pilot plant and it was positively assessed. However, this technique has been withdrawn from the market

for technical reasons (it is too sensitive a technique and tends to break down frequently).

- Technique 3 (mechanical dismantling, sorting and recycling of the different materials, including the neutralisation of CFCs using the Solvay-Method or through high temperature incineration as well as the re-use of polyurethane (PUR) foams) is available and can be regarded as ecologically acceptable.

6.2 Other related innovations

The *main innovation* of this case study – thus the innovation on which the analysis is centred – is the above mentioned recycling procedure that closes material cycles, especially of CFCs, in an ecological way. At WHITE Ltd. this innovation was accompanied by other changes in the technical and organisational framework. Two *institutional arrangements* should be explicitly mentioned here, because they may have direct and indirect effects on employment:

- The co-operation between WHITE Ltd. and enterprises offering wage-subsidised jobs for unemployed persons (e.g. BLACK Ltd., a non-profit make-work and training scheme). This was the main subject in the interviews with the managers of those firms.
- Furthermore, WHITE Ltd. became a member of "RAL-Gütegemeinschaft", an association of firms that have obtained a RAL-Label for recycling refrigerators according to the ecological criteria set by RAL.

7 The company adopting the innovation: WHITE Ltd.

WHITE Ltd. was founded in 1987 and explicitly aimed at linking two goals: achieving the maximum operating result by recovering as much CFC as possible from refrigerators. In fact, the signing of the Montreal Convention (that laid down deadlines for the banning of CFCs) was pushed to a large extent by the strong public awareness of CFC-issues. This is also the reason why no intervention from the federal government was needed to enforce the separate collection of refrigerators at the local level: By 1988 almost all municipalities in Germany had on their own initiative established such a recovery system for end-of-life refrigerators from households, the costs of which were covered by raising local waste management charges. Therefore, there was an increasing demand for appropriate recycling plants.

WHITE Ltd. seized the opportunity to penetrate the emerging niche market with an innovative process. The first stage of recycling (extracting CFCs from the cooling aggregate) was already being done in some small recycling firms. They used a simple manual technique unsuitable for industrial application. The first task WHITE Ltd. faced was to implement R&D measures to adapt a traditional technique to industrial constraints (investment in mechanical engineering). From 1988 on, the refrigerators the municipalities collected were delivered to WHITE Ltd. and recycled with the help of this process. However, it soon became clear that two environmental problems had to be solved:

1. a logistical problem. The refrigerators often arrived in a bad state, losing up to 50% of their CFCs during collection and transport;
2. a suitable technique for the recovery of CFCs from insulating foams was not yet available but, in fact, a refrigerator contains more CFCs in its foams than in the cooling aggregate.

Attempts were made to optimise the process, and in 1990 further R&D measures led to the introduction of the innovation at WHITE Ltd. that constitutes the *main innovation* of this case study. It bears the following characteristics:

- The recovery of CFCs is mainly done with mobile plants to avoid CFCs escaping during transport. These mobile plants are installed on the trucks that go to the collecting points and do the complete recycling on the spot.
- Before the refrigerator is dismantled, the cooling agents CFC-R 12, R 22 etc. and oil from the condenser are recovered with the help of a suction system. The CFCs are filled in special containers and sent to Solvay for neutralisation at temperatures of above 1,300° C. The used oils from compressors go to oil refineries for treatment.
- The next step consists in manually pre-dismantling the refrigerator, in the course of which some reusable parts are separated and treated for specific re-use or recycling (like glass, the compressor, and electric cables, etc).
- Then the refrigerator cabinets are shredded mechanically. Fe- and NFe-metals as well as insulating foams are separated and treated for re-use. The insulating foams are transformed through a cellular degassing process into CFC-free PUR-powder which is used as bonding agent in other industries. The recovered CFC-R 11 is sent to Solvay for treatment. The metal and plastic parts are shipped to recycling enterprises.

7.1 Development after the introduction of the innovation

In 1993 the technical performance of WHITE Ltd.'s innovation was improved with the introduction of a computer-based control system with which the technique was optimised. Further technical innovation would only lead to incremental progress.

At that time, holders of end-of-life refrigerators paid € 20-40 for a refrigerator to be treated. Quality was related to the amount of CFCs that was recovered and was an important criterion at that time, that is why WHITE Ltd. was successful with its eco-innovation, though it asked the highest prices in the market. It increased its capacity to 12 plants in stage 1 (each with a capacity of 80,000 refrigerators a year, some of them operating in other European countries) and 4 mobile plants in stage 2 (each with a capacity of 120,000 refrigerators a year).

In the first phase (1988-1994) WHITE Ltd. benefited from its pioneer position and captured about 10% of the market volume by introducing an innovation that satisfied the great demand in the market. However, between 1994 and 1997, the business went through a serious crisis. Pressure on the waste treatment market grew, especially through the penetration of power-supply firms, which pushed incineration as a less expensive solution for waste treatment.

Though the technical standards of WHITE Ltd. still represent the BAT-Standard (Best Available Technology), the business trend was negative for several years. Finally, in 1997 the management managed to stop the decline with a successful PR campaign targeting municipalities, showing them the ecological merits of WHITE's technique. In 1998, WHITE Ltd. obtained the RAL-Label. Together with other RAL-labelled enterprises (currently 8 firms), it built an association that aims to enforce the establishment of ecological standards in the market. Then the situation at WHITE Ltd. stabilised. But currently the discussion about a law on electrical and electronic scrap is creating a feeling of uncertainty again. In fact, in the last few years, strong competition has led to a broad price range directly related to the different types of treatment: at present the price for the treatment of a refrigerator lies between € 6 and € 25 (this is the price WHITE Ltd. and all those firms charge that ensure the recovery and, as far as possible, separate closed material cycles for all materials including CFCs).

Firms with high environmental standards are therefore afraid that they will be faced with the competition of ecologically less efficient firms, if legislation does not lay down high ecological standards for the recycling of refrigerators. But neither the draft of the German government nor the proposal for an EU-directive on electrical and electronic scrap establish thresholds for CFC recovery. Thus, WHITE Ltd. supports the voluntary RAL-labelling and by means of intensive advertising tries to convince as many actors as possible of the label's advantages. These actions of the RAL-Gütegemeinschaft aiming at enforcing high standards for refrigerator recycling can be seen as a political measure that belongs to a new generation of instruments intended to achieve environmental goals independently of government intervention.

While WHITE Ltd. has obviously had trouble keeping its position in the German market, it has successfully expanded its activities in other European countries. Currently it has 17 plants for stage 1 (6 in Germany) and 3 plants for stage 2 operating in several European countries, especially in Scandinavia.

7.2 Employment effects of the innovation

7.2.1 The functional equivalence approach (FEA)

Three Functional Equivalence Approaches (FEA alternatives) were proposed:

1. to compare the situation before and after the innovation (FEA-1),
2. to identify a similar context where this innovation did not take place (FEA-2),
3. to identify a theoretical reference (not applicable in this case).

To assess the employment effects of the innovation at WHITE Ltd., for pragmatic reasons, particularly because of the availability of reliable data, it seemed appropriate to choose FEA-1 first. Thus, the evolution of the innovating firm was observed over a period of about 10 years after the introduction of the innovation. The advantage of this approach is that it provides information on dynamic processes on the microeconomic level.

However, regarding the results of this first approach, the question arises whether the historical evolution of the innovating firm has essentially been any different from that of its competitors. This question is pertinent, for if the evolution of innovators and non-innovators in the whole sector has been similar, this may indicate that possible employment effects discovered in this study do not actually result from innovation, but from other factors that influenced all the firms in the sector similarly during that period (legislation, international competition, etc.). As information on a sectoral level is not reliable, it seemed reasonable to complete the analysis of the case study using FEA-2, i.e. comparing the evolution of employment in the innovating firm (and respective value chain) with that of non-innovative firms in the sector. It was easy to use this approach, because the only aspect that is essentially different in the method of WHITE Ltd. is the innovative technique of recovery and treatment of CFCs. To that end, data on firms working with other technical alternatives were also collected. In the refrigerator recycling industry there are three types of firms:

- "shady" actors, that do not really recycle the end-of-life products (i.e. they export them or dispose of them illegally). The current data about the number of refrigerators that are actually recycled in Germany point to an expansion of the "shady" sub-sector in the last 10 years. Most firms operating in the EES recycling are very small enterprises (1-2 persons). They work using only a simple transport and storage infrastructure. The direct employment effect is normally reduced to the employment of the owner himself and his relatives. However, they should be regarded as competitors that changed the market structure by transferring parts of the value chain to other countries, and that have thus probably had an important impact on employment in the whole recycling industry;

- firms performing the first stage of recycling (extraction of the CFCs from the cooling aggregate), shred the refrigerator cabinets, separate the recyclable materials (metals, plastics) and incinerate all the rest (including foams) in waste incinerators. This alternative is ecologically questionable, for it is uncertain whether the CFCs from the foams have been eliminated. But these firms work within the legal framework (which makes them respectable) and they enjoy comparative cost advantages, because they use a less demanding technique for the second stage of recycling. However, the employment situation at these firms does not significantly differ from that at WHITE Ltd., for the labour-intensive first stage of recycling is common to both WHITE Ltd. and these firms;
- firms that have a similar value chain as WHITE Ltd., but send the separated CFC materials to special plants for incineration. As incineration in special plants is seen by experts as an ecologically acceptable alternative, from an ecological standpoint they may be regarded as direct competitors to WHITE Ltd. However, we may assume that the cost situation of these firms is quite similar to that of WHITE Ltd. In fact, though special incineration is much cheaper than the Solvay cleavage method[6], WHITE Ltd. believes that this particular cost item is not decisive, for the amount of CFCs recovered for treatment is relatively small. These higher costs are seen as the price for the qualitative consistency of the method of WHITE Ltd., which is based on the principle of closing material cycles, including CFC cycles. Since the latter is a strong selling point, the costs are accepted.

Though it is impossible to come up with quantitative statements, these remarks may provide a better basis for the interpretation of the results. In fact, firms offering services at much lower prices in the German market normally do not observe appropriate ecological quality standards for refrigerator recycling. Lower costs result a) from saving labour in the first stage of recycling in Germany (no manual dismantling and recovery of reusable parts) and/or b) from lower costs in the second stage, normally by incinerating refrigerator cabinets, including CFC-foams, in "normal" waste incinerators (probably including the recovery of some metal content).

7.2.2 Evolution of employment in the innovating firm

In 1987/1988, an engineering company was asked to plan and assist in the construction of the recycling plant at the company headquarters. Four experts from the consulting firm worked full-time on this project. With 15 skilled employees

---

6 The average costs for the recovery, transport and disposal of the CFCs of a single refrigerator are currently estimated at € 5. The disposal of CFCs at Solvay S.A. costs € 3,000 a ton, incineration in a "normal" waste incinerator costs about € 100-400 a ton and incineration in a special plant for hazardous waste about € 500.

from WHITE Ltd. they built the first plant and improved it over the next few years. Currently, the number of employees is much smaller: in the consulting firm only one engineer is still working mainly for WHITE Ltd. At WHITE Ltd. itself, five skilled employees are charged with the maintenance and repair of the equipment. The persons working in the mobile plants have been trained to solve urgent technical problems while the plant is operating away from the company headquarters.

In 1994, the firm had its largest workforce ever: 120 employees. Three skilled workers worked in each mobile plant.

During the crisis in the years 1994-1997, WHITE Ltd. had to lay off 60 persons. The difficult situation the company was in required changes in the workforce. With the introduction of a computer-based control system in 1993, the number of employees in each mobile plant decreased from three to two. Furthermore, a trend toward regional reorganisation allowed a reduction in labour costs, especially in the first stage of recycling. Additionally, an expansion into foreign markets compensated for the negative trend in Germany.

In 1997, 15 stationary plants were installed in recycling firms that have taken over the manual dismantling and the recovery of CFCs and oil from cooling aggregates with equipment from WHITE Ltd. As mentioned above, these enterprises mostly employ poorly-skilled, long-term unemployed persons that qualify for publicly subsidised wages. Since 1994, there have only been direct, positive employment effects in this field. One of these enterprises is BLACK Ltd. The co-operation of WHITE Ltd. with firms like BLACK Ltd. allowed steps of the value chain of refrigerator recycling to be outsourced. At this point, it seems appropriate to describe the value chain of refrigerator recycling.

7.2.3 Value chain method: setting the system boundaries

Figure 3 shows the simplified material chain of refrigerator recycling. It follows the LCA methodology that makes a systemic distinction between stock (◊) and transforming (□) items to ensure methodological stringency. Stock items are steps in the material chain, transforming items are steps in the actual value chain. In fact, only at transforming steps is it possible to change the value of materials, because that is where they are processed with the help of workers, machines, energy, etc. Thus the □-chain in the diagram may be seen as a value chain in which:

- the starting point is an end-of-life refrigerator in a household or company (first boundary)
- the relationship between steps in the chain and employment easily becomes evident, for employment is mainly related to the activities at transforming steps. For instance, when the municipal services pick up refrigerators from households, there will be employment effects. And vice versa: when owners

take their end-of-life refrigerators to a recycling centre, no jobs will be created in the field of transportation.

- the system boundaries become evident, too: the black lines represent the chain that is the object of observation; the grey lines show the excluded technical paths, for instance "disposal in landfill".

When it comes to employment effects, the grey lines play a particularly important role in this diagram. They represent paths that are excluded from the material chain, because they are not part of the recycling concept of WHITE Ltd. Therefore, they nicely show the "alternative" chains, where jobs were probably lost due to the innovation.

In this context, some additional comments have to be made:

- first, a period of 10 years was set for the assessment of employment effects, though for several aspects information was not available for the whole period;
- second, in most studies it is normally assumed that system boundaries correspond to the existing political boundaries; however, this assumption is not methodologically stringent. In fact, from an economic point of view, if jobs are killed in Germany because the same jobs are created in France, it is not possible to say whether the innovation inducing this change has had a negative or positive net employment effects. Thus, whenever positive (negative) net employment effects of innovations are mentioned in this study, this just means that some jobs have been created (have been lost) in Germany, specifically within the value chain during the period observed;
- and third, last but not least, it is important to be aware of a fundamental limitation of this method, namely that with a vertical value chain approach we will not be able to assess (horizontal) effects produced by competitors.

7.2.4 Upstream employment effects

In this case study, upstream from the innovation the most labour-intensive step in the value chain is manual dismantling. As mentioned above, a typical development in the employment situation in the recycling industry and especially in the EES recycling sector, has been the creation of numerous jobs for long-term unemployed, unskilled and handicapped persons in this area.

Black Ltd. is a subsidized, non-profit organisation that aims to integrate long-term unemployed persons in the labour market. It has a training centre for waste treatment where 18 unemployed persons can obtain a certificate as skilled workers for the recycling of electronic scrap. These training places are directly and indirectly subsidised by the public authorities. The direct support is provided by financing ABM (the national scheme for job-creating measures), the indirect support is given by demanding services and goods produced by BLACK Ltd.

In 1995 the regional administration invited tenders for the recovery and recycling of refrigerators and BLACK Ltd. was awarded the contract. For each refrigerator, it is paid € 20. About 7,000 refrigerators are recovered annually by BLACK Ltd. The first stage of CFC recovery is done at the company headquarters with equipment from WHITE Ltd. The firm was checked by RAL inspectors as part of the chain assessment during the labelling of WHITE Ltd.

Two full-time jobs at BLACK Ltd. consist in dismantling the refrigerators, extracting the CFCs and separating the reusable and recyclable parts. The refrigerator cabinets are then sent to WHITE Ltd. for the second stage of recycling. The average labour input for one refrigerator at BLACK Ltd., including the transport to WHITE Ltd., is about 20 minutes, i.e. approx. € 2. This only works because of the low wages that these employers pay (on average a married worker with one child gets € 1,300 a month).When the firm was founded, the management of BLACK Ltd. was aware of the need for a sustainable development in the region. Thus, they tried to create job opportunities mainly related to environmental protection. Recycling with high ecological standards became one of the firm's main fields of activity. This led the management of BLACK Ltd. to WHITE Ltd., which was known for its high ecological standards. BLACK Ltd. contracted it to do the second stage of recycling.

Both firms underlined in interviews that they enjoy mutual economic benefits from their co-operation. In the joint interview it became clear that the relationship between the managers of the two firms is not just a business relationship; they also agreed on the evaluation of current political issues and both argued against the new ordinance and its inherent threat to jobs. In their opinion, changes expected in the legal situation may have negative effects on both firms. If the proposed legislation on EES is actually passed, the survival of BLACK Ltd. – and of all the firms in the business of collecting and manually dismantling refrigerators – is at risk. These firms fear that producers of electrical and electronic equipment will not use the existing regional networks for recovery. For WHITE Ltd. the first stage of recycling in its present form can only be kept in Germany, if labour costs are very low (i.e. subsidised, like at BLACK Ltd.). That is why they called for an ordinance that aims at achieving sustainable regional development goals and takes into account the existing networks between firms offering high ecological recycling standards and publicly subsidised firms that take on the labour-intensive part of the activities

Figure 3: Value chain of refrigerator recycling

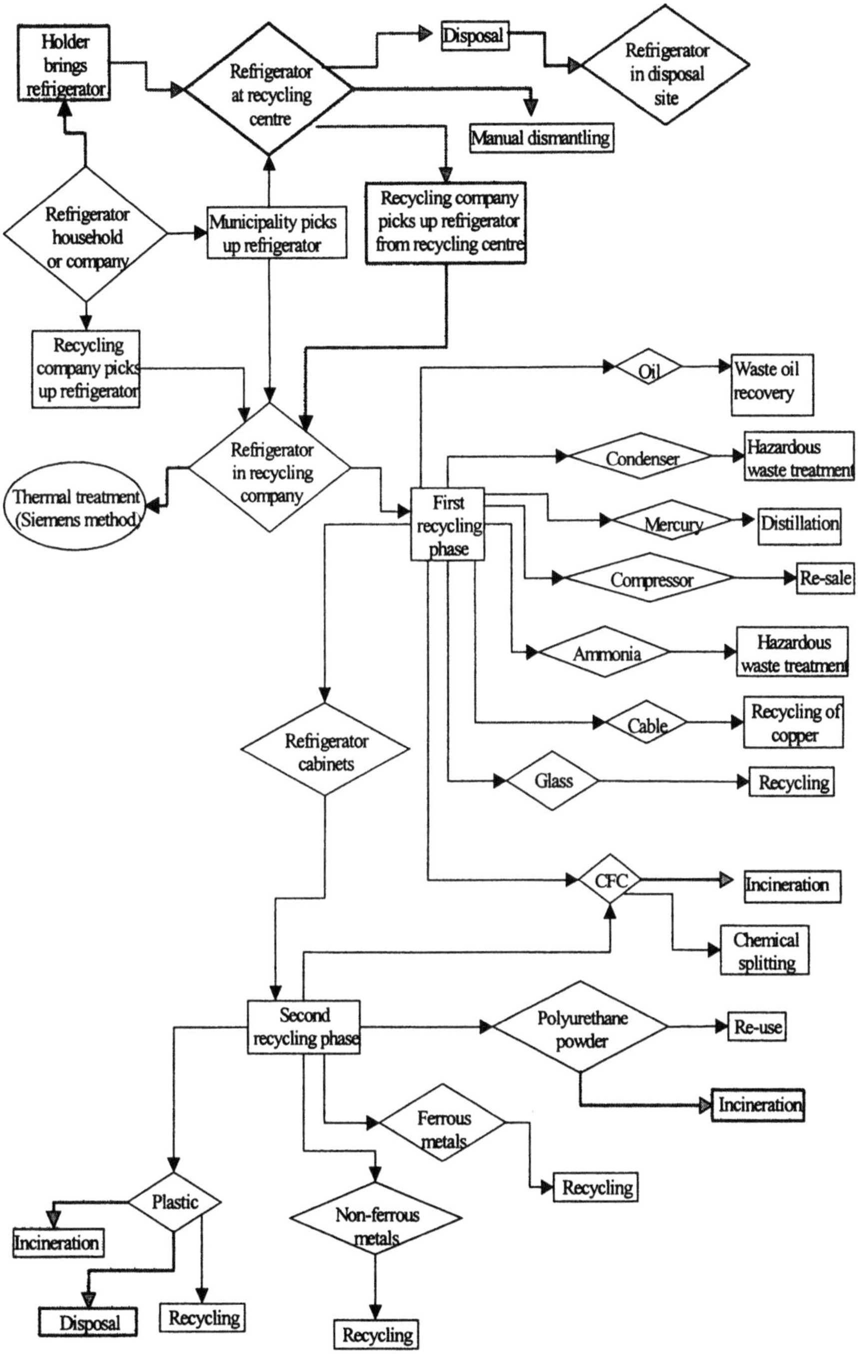

If we take the employment effects at BLACK Ltd. as a basis of calculation (two full-time jobs for 7,000 refrigerators), the innovation of WHITE Ltd. secures about 28.5 jobs in the first stage (currently WHITE Ltd. treats 100,000 refrigerators a year). An extrapolation for the whole market segment (assuming that 600,000 refrigerators go through the first stage of recycling in Germany) shows that upstream 171 jobs would be directly related to this innovation if all firms were working like WHITE Ltd. and BLACK Ltd.

7.2.5 Downstream employment effects

It is difficult to quantify the employment effects for the downstream chain, because the quantities of materials delivered by WHITE Ltd. to other firms are relatively small, that is why in the opinion of all the interview partners their employment impact is insignificant. The value chain is made up of the following paths, which were mainly evaluated with FEA-2:

- **Waste oil recovery** – there is a binding legal framework for the treatment of used oils, thus jobs related to this path are unrelated to the innovation of WHITE Ltd. All firms that recycle refrigerators have to have their used oils treated in special plants. Currently the price for treatment is about € 400-500 a ton including transport and logistical costs. Therefore it is an important cost item for recycling firms. To comply with legal provisions, delivery firms have to organise the transport. It is thus probable that additional jobs created through recycling on account of the need to conform with the law do not emerge in oil refineries, but in the recycling industry itself.

- **Hazardous waste treatment** (condenser with PCB, ammonia) – The comments on waste oil also apply to hazardous waste. Prices of chemical-physical treatment and disposal vary, but the quantities delivered by WHITE Ltd. are small (only 3% of the refrigerators use ammonia, and only one in a thousand has PCB in its condenser). Thus this treatment does not have any impact on employment.

- **Distillation of mercury** – only very old refrigerators have a mercury switch. They are sent to a special plant for recovery. Costs and jobs are a quantité négligeable for WHITE Ltd. No employment impacts are expected on the following steps in the value chain.

- **Resale of (reusable) compressors** – there is a great demand for used compressors. Intermediate traders deal in these end-of-life products, which are partly used in workshops to repair electrical equipment. In Germany, the average market price of used compressors is about € 1.50 per unit. All firms in the refrigerator recycling business have an interest in trading in compressors. Thus, no matter what form of recycling they use, innovative and non-innovative firms will have the same impacts on the downstream value chain.

- **Fe-metal recycling**: Compressors that cannot be re-used go to metal-recycling enterprises together with the metal cabinets to be recycled. The comments on the resale of compressors also apply to this path. Fe-metals constitute the most important material content of a refrigerator (accounting for 40% - 70% of its weight) and there is a great demand for these secondary raw materials with prices between € 40 and € 70 per ton. Prices are currently going up. Consequently, the recovery of these materials is one of the most important motivations of all the EES recycling firms. WHITE Ltd. probably derives some comparative advantages from its main innovation, because the Fe-metals it recovers following the mechanical shredding procedure is fairly unadulterated and can therefore be sold for a good price. However, from an employment perspective this may have negative impacts on the job situation downstream. Materials that are not sorted that well normally require additional treatment in shredding firms (and add more steps to the value chain). Two metal-recycling firms (YELLOW Ltd. and BLUE Ltd.) were interviewed. Price uncertainty and concentration trends in the market are threatening SMEs (most of them have fewer than 40 employees). The Asian crisis last year had a strong impact on the scrap market. In Germany, the steel industry, the buyer of sorted Fe-metal scrap, negotiates the prices with scrap traders every month, so fluctuation is a constant problem. Scrap recycling firms claim that the quantities of scrap are decreasing in Germany and that the prices they have to pay to suppliers are increasing, whereas the steel industry is highly dependent on trends in the international market. Nevertheless, both interview partners believe that scrap recycling is a promising business. They do not see any need for additional capacities in Germany. However, to remain competitive in the market SMEs have to cut costs and an optimisation is only possible if labour costs are reduced – in other words, jobs will be lost during the next few years.

- **Non-Fe-metal recycling** – The comments on Fe-metals also hold true for Non-Fe-metals, especially for aluminium (copper only accounts for 3% of this group of metals; electric cables are delivered free of charge). Aluminium is a secondary raw material which is in great demand, with prices varying between € 300 and € 600 a ton.

- **Plastic recycling** – The plastic scrap normally goes to plastic waste traders. One of those firms, GREEN Ltd., was interviewed. GREEN Ltd.'s representative estimates that 5,000-6,000 firms operate in this sector. GREEN Ltd. accepts plastics of different sorting qualities, the best sorted plastics (like those of WHITE Ltd.) go to plastic manufacturers, where jobs have been created in the last decades.[7] The specific impact of WHITE Ltd. on employment is however not significant.

---

7 However, 40%-50% of the waste plastics are disposed of by incineration.

- **CFC chemical cleavage** – This substance goes to an intermediary firm, RED Ltd., which is charged with collecting all CFCs that are treated at Solvay S.A. by being chemically split. Solvay is the only firm in the world that uses this process. RED Ltd. is responsible for the pre-treatment of CFCs (distillation) and the transport (in pressure vessels). The most important customers are the Netherlands, for all the CFCs from the Netherlands are sent to Solvay S.A., and Austria. The interview partner believes that two thirds of all CFCs in Germany go to Solvay S.A., too, and one third is burnt in special waste incinerators. Currently, 30 persons work for RED Ltd., but only 3 of them in the field of CFC treatment, thus the impact of WHITE Ltd. is estimated to be very small.
- **CFC-free PUR-powder** – The CFC-free PUR-powder is sold as a binder to firms working with oils. The powder is used for cleaning activities in the firms. This path does not create any jobs.

8 Conclusions and discussion

As mentioned above, this case study aims at achieving two main goals: first, it is meant to clarify the motivations and decisions of individual innovative firms and provide information about the impacts of a specific eco-innovation on employment. Second, it is designed to be used as an explorative analysis to test the adequacy and limits of the methodology employed (value chain method) for the purpose of this study.

8.1 Employment effects

Table 4 gives a summary of the most important findings of this study. The first column shows the value chain of the recycling process, the second the firms that were interviewed on each item. The overview of the employment effects along the value chain assessed with two different methods, FEA-1 and FEA-2, constitutes the core part of the table (third and fourth columns). With the first method we assessed the employment in all steps of the value chain after the introduction of the innovation at WHITE Ltd. The situation before the innovation was introduced served as the benchmark. With the second method we assessed the possible employment situation in the different stages of the value chain, if WHITE Ltd. had not introduced its eco-innovation. Thus, in this case the benchmark is the employment situation (ex ante, ex post) at WHITE Ltd. itself and at its competitors (those, who use a legally acceptable, but less ecologically effective recycling technology). Actually, only stages of the value chain that changed because of the innovation (marked with "yes" in the FEA-2 column) are expected to show any effects which are mainly or exclusively induced by the innovation. The comments in the last column give some additional information that is useful for the classification of the net employment effects (direct, indirect effects).

By comparing the results of FEA-1 and FEA-2, we are able to qualitatively distinguish net employment effects that can be put down to the eco-innovation from effects that are attributable to other causes. Those effects are related:

a) to R&D activities and to the employees needed for the construction and maintenance of the mobile and stationary plants of WHITE Ltd.,

b) to the jobs in the first stage of recycling in the innovating firm and in the upstream chain. Actually, most of the jobs related to the main innovation addressed in this study were created upstream,

c) to the few jobs in the second stage of recycling (recovery of CFCs from insulating foams) and the production of CFC-free PUR-powder in the innovating firm.

Table 4: Net employment effects in the value chain assessed with FEA-1 and FEA-2

Value chain	Interview partners	Employment effects: situation before and after the innovation (FEA-1)	Employment effects compared with non-innovative alternatives* (FEA-2)	Comments
Upstream chain				
Transport of one refrigerator from household to recycling centre	WHITE Ltd. BLACK Ltd.	+/-	no	Jobs are shifted from municipal services to private enterprises
Manual pre-dismantling	WHITE Ltd. BLACK Ltd. ZAUG GOAB	+	no	Since 1994 jobs have shifted from WHITE Ltd. to non-profit enterprises
Internal chain				
Construction and maintenance of plants	WHITE Ltd.	+	yes	Has been decreasing since 1990
First stage of recycling (recovery of CFC-cooling agents)	WHITE Ltd. BLACK Ltd.	+	yes	Partial shift of jobs from WHITE Ltd. to other firms
Second stage of recycling (recovery of CFCs from insulating foams)	WHITE Ltd.	+	yes	

Table 4: continued

Downstream chain				
CFCs	RED Ltd.	+/-	yes	
Oil	WHITE Ltd.	+/-	no	
Condenser	WHITE Ltd.	+/-	no	
Mercury	WHITE Ltd.	+/-	no	
Compressor	WHITE Ltd. BLACK Ltd.	+	no	is sold
Ammonia	WHITE Ltd.	+/-	no	
Cables	WHITE Ltd. YELLOW Ltd.	+/-	no	
Glass	WHITE Ltd.	+/-	no	
Fe-metals	BLUE Ltd	+/-	no	are sold
non-Fe-metals	YELLOW Ltd.	+/-	no	are sold
Plastics	GREEN Ltd.	+/-	no	are sold
PUR-powder	WHITE Ltd.	+	yes	is sold

Remark: * Pre-dismantling, shredding, separation of recyclable materials and incineration of the residual cabinets and foams.

In sum, the following conclusions can be drawn:

- In Germany since 1987 most municipalities have established facilities to pick up end-of-life refrigerators from households. Sometimes they contracted private recycling centres for this job. Jobs in the upstream chain collecting refrigerators were thus created before WHITE Ltd. introduced its innovation. These positive direct employment effects were induced by a public awareness of environmental issues in the late 1980s (creation of demand). The net effects of the innovation are zero. As the number and quality of those jobs do not depend on whether the firm is publicly or privately owned, there have not been any substantial changes in employment regarding collection.
- Manual dismantling was a common technique in the early 1980s in Germany, but the diffusion of mechanical methods made some of those jobs redundant (indirect negative effects of technology change). Only pre-dismantling still offers job opportunities for unskilled labour. WHITE Ltd.'s innovation created some jobs in the firm itself and in the upstream value chain (direct effects), but

these jobs replaced jobs in non-innovative firms – thus on a larger scale the net effects are zero.

- The main innovation of WHITE Ltd. was induced by environmental policy measures starting in 1987 (banning of CFCs). Municipalities had to find an ecological solution for end-of-life refrigerators. For this reason waste charges were raised and a new market was created, in which SME pioneers like WHITE Ltd. were able to seize the opportunity. The direct employment effects in the innovative firm were positive (R&D, construction). The jobs were for skilled persons. However, once the innovation had been implemented, the number of employees working in those areas decreased significantly.

- In 1989 the German government announced an ordinance for electronic waste (including refrigerators) enforcing recycling activities and energy recovery. As a consequence, large capacities for the recycling of EES were created in Germany. Strong competitors, especially from the energy industry, penetrated the waste management sector and pushed incineration (that is why they were chosen as a benchmark for the FEA-2 assessment). However, the expectations concerning the quantities of EES available for recycling have not been completely met. In the course of time, the competitiveness of environmental options for the disposal of CFCs has declined compared with the incineration option. WHITE Ltd. laid off workers. This was a negative indirect effect caused by market competition

- In the following years WHITE Ltd. tackled the crisis by introducing further institutional and technical innovations, for example:

- the first stage of the recycling process partially moved to BLACK Ltd. and other socially committed firms where unskilled labour is supported through public subsidies. This shift of jobs was due, on one hand, to WHITE's cost reduction strategy and, on the other hand, to political concern about unemployment. The shift of jobs had positive direct net effects on the upstream chain, because presumably more (unskilled) workers found a job in those enterprises than WHITE Ltd. had previously employed. Thus, compared with those of competitors, this value chain is probably more labour-intensive than the alternatives.

- The diffusion of mechanical dismantling processes (by WHITE Ltd. and its competitors) replaced all of the manual dismantling that used to be the main technique in Germany in the 1980s. Now there are not any firms in Germany that still apply the old technique. At first glance, the elimination of this path of the value chain might induce negative employment effects in Germany, mainly with regard to unskilled jobs. But actually, it seems that this recycling path has not been totally eliminated through technical innovations. In particular recycling alternatives like the one WHITE Ltd. came up with, which requires reusable parts of refrigerators (for example, the compressors) to be carefully separated from the cabinets, involve a considerable input of labour. Experts think that the decrease in the number of jobs in this field has not been dramatic so far. The refrigerator recycling industry is now confronted with increasing exports of end-of-life products to Eastern Europe, where recycling consists of manual dismantling and re-use of usable parts. This development suggests a transfer of parts of the value chain from Germany to other countries. In this study, these effects have not been assessed, because no reliable information was available. However, we may assume that there are probably positive direct and indirect employment effects outside the system boundaries.

- Two mobile CFC-extraction plants expanded the geographical dispersion of WHITE Ltd.'s activities to other countries (about 4 skilled workers work at those plants; other effects of this expansion have probably been felt at a local level abroad, i.e. outside the system boundaries of this case study, too).

- Intensive marketing and advertising of the ecological advantages the two-stage recycling system offers. Since 1997 no worker has been made redundant. This can probably be classified as an indirect effect of the innovation, for it is a result of marketing strategies which increased the acceptance of WHITE Ltd.'s higher prices (because of its higher ecological standards) on the part of its customers.

- So far, the recovery and pre-dismantling have normally been organised in labour-intensive, regional networks in which jobs for unemployed, unskilled persons have been created with public subsidies. Currently, the ordinance for electronic waste is again the object of political discussion. According to the government's draft, producers will probably be forced to recover their own end-of-life products and recycle them. An ordinance similar to the one on the Dual System for packaging waste is now under discussion. If such an ordinance is passed, it is expected that the recovery and pre-dismantling will be concentrated in as few as 5 big recovery locations in Germany.

- The employment effects in the downstream chain are less significant than in the upstream chain. The innovation of WHITE Ltd. mainly affects the recovery and treatment of CFCs. Therefore, employment only changes at the steps directly related to the recovery and treatment of this hazardous substance. Apart from the jobs created at WHITE Ltd itself, the innovation has hardly generated any jobs downstream, because either the quantities for recycling are relatively small or the corresponding paths in the value chain are not labour-intensive. However, the importance of the downstream chain from a business standpoint should not be underestimated. The very business objective of the firm is to recover plastics and metals and sell these secondary raw materials in a market with increasing potential.

8.2 Some remarks about the methodology

As part of IMPRESS, the case studies are expected to contribute to a methodological improvement, too. As for this goal, the following comments should be made.

- Theoretically, employment effects in competitive markets may be:
- related to their location, i.e. in the firm we are dealing with direct effects, outside the firm with indirect effects. Theoretically, this definition is not suited to facilitate an understanding of the complexity of the issues at stake;
- related to their causes (motivation-oriented assessment): if employment results from intentional cost reduction or sales expansion strategies, this is a case of direct effects. But employment effects may also result from a reaction to changes occurring outside the firm, for example to changes in legal provisions, or to changes caused by the competition in the market, and in particular from reactions to changes brought about by substitution mechanisms and/or income effects – these are indirect employment effects;
- related to a particular change (i.e. a specific innovation) no matter where and why the effects occur within given boundaries.

The value chain approach used in this case study falls into this last category. One of the most important results is that this method makes transparent the role of the system boundaries for the assessment of net effects. In order to distinguish between direct and indirect effects, a motivation-oriented assessment was used. Because the linkage of the value chain approach with the motivation-oriented approach is not trivial, some critical comments should be made:

- The assessment of the *direct employment effects* of an innovation along the value chain is difficult, because direct effects result from *intentional* measures at the company level. As said above, if the innovation is introduced to reduce labour costs and some jobs actually become redundant, these are direct effects. However, if the innovation is introduced because legal provisions or market pressure force the firm to do this, the employment effects triggered by the innovation are indirect effects. But when firms in the value chain modify their own production or organisation in order to adapt to the innovative measures taken by suppliers (or customers) it is debatable whether the employment effects of these measures should be seen as direct or indirect effects.
- As for *indirect labour effects*, other problems arise. Indirect effects are caused by changes concerning the market position of the firm (thus they are not directly related to the innovation, but to competition, marketing strategies, institutional arrangements, legislation, etc.). If there is an increase in income (profits or wages) due to the innovation and this increase makes it possible to create some jobs, probably in other industries, these effects are indirect effects of the

innovation. However, an assessment of this type of effects is not possible if one only looks at the value chain. Substitution effects by competitors cannot be properly analysed with this approach either. In fact, the incompatibility of both perspectives (motivation-oriented and value chain) becomes evident when an innovation prompts job losses in other industries. From a motivation-oriented perspective, these effects are indirect effects caused by changes in the competition in the market. From the point of view of the value chain method, however, these effects are not normally taken into account, because they occur outside the system boundaries. But if the system boundaries are wide enough to include them, they should probably be directly attributed to the *main innovation* and therefore be seen as direct effects.

The present study shows that the distinction between direct and indirect effects requires a consistent approach. By using a value chain approach, especial attention should be given to the definition of the system boundaries. The use of a two steps assessment method (FEA-1 and FEA-2) seems to be a suitable way to make the value chain method compatible with a motivation-oriented assessment.

8.3 Final conclusions

A case study is per definition a descriptive tool. It does not allow general statements about the causalities of the elements observed. Moreover, the results of a case study cannot simply be extrapolated. However this case study is expected to deliver some empirical basis for a better understanding of the relationship between employment and recycling in general. The following assessment sums up the results of the present case study:

- In the short term, the innovation had positive employment effects on the firm and the upstream value chain (but probably negative indirect effects on the sector). In the medium term we identified negative effects in the innovative enterprise itself. The unfavourable market situation created incentives for the innovative firm to make new innovation efforts. These had no positive employment effects on the enterprise, but they stopped the downturn and had positive (direct/indirect) effects on the upstream value chain. This positive effects in the upstream chain would not be so significant, if the concerned jobs were not publicly subsidised. The long-term direct employment effects of the recycling innovation seem to be negative for the innovative enterprise in this specific case study, positive for the upstream chain and insignificant for the downstream chain.
- In spite of the hopeful message that we quoted in the first part of this chapter, the EU-Commission also recognises the problem that inevitably arises from the high expectations of a "double dividend" from recycling, as the following text shows: "The implementation of recycling objectives in the context of an environmental policy has given rise to a situation where the activity of recy-

cling is not profitable, unless some direct or indirect public intervention takes place" (European Commission, 1998: 2). This case study bears out this statement. Jobs in the recycling sector in Germany are highly dependent on public support.

- The Commission's report also states that "if markets function correctly and in conditions of maximum efficiency and minimal costs, recycling may become profitable in an increasing number of cases" (European Commission, 1998: 2). It remains questionable, whether the best technical option of recycling will succeed in the European market, if legal regulation does not stipulate minimal ecological standards on the international level.

Interview partners

- Ifeu-Institut, Heidelberg (Jürgen Giegriech and Horst Fehrenbach, Researchers).
- WHITE Ltd.- refrigerator recycling firm, 60 employees, interview partner: sales manager.
- BLACK Ltd. – publicly supported, non-profit enterprise for the creation of employment, 30 employees, 60 training places, interview partner: manager.
- RED Ltd. – small firm (30 employees) for the collection and pretreatment of CFC-substances; interview partner: head of department.
- ZAUG – Zentrum Arbeit und Umweltschutz, Gießen (interview partner: head of department).
- GOAB – Gemeinnützige Gesellschaft Offenbacher Ausbildung und Beschäftigung (interview partner: head of department).
- GREEN Ltd. – small trading firm (3 employees) for waste plastics, interview partner: proprietor.
- BLUE Ltd. – small ferrous metal recycling firm (30 employees), interview partner: proprietor.
- YELLOW Ltd. – metal-recycling firm (fewer than 40 employees), interview partner: proprietor.

References

Bundesverband Sekundärrohstoffe und Entsorgung (1998), *Elektronikschrottrecycling – Fakten, Zahlen und Verfahren*, Bonn.

European Commission DG III – Industry (1998), *The Competitiveness of the Recycling Industries*, COM 463 final, Brussels.

Hafkesbrink, J., M. Halstrick-Schwenk and K. Löbbe (1999), Abschätzung der innovativen Wirkungen umweltpolitischer Instrumente in den Stoffströmen Elektroaltgeräte/ Elektronikschrott, in: P. Klemmer (Ed.), *Innovationen und Umwelt*, 167-201, Berlin.

Halstrick-Schwenk, M. (1999), Entsorgung und Wiederverwertung von Elektroaltgeräten und Elektronikschrott – Ökologisches Gefährdungspotential und umweltpolitischer Handlungsbedarf, in: *RWI-Mitteilungen* Vol. 50, No. 1-2, 81-107.

Ifeu (1997), *Ergebnisbericht zum UBA-Forschungsvorhaben Nr. 10310606 "Ökologische Bilanzen in der Abfallwirtschaft", Fallbeispiel: Verwertung von Haushaltskühlgeräten*, Heidelberg.

Stiftung Arbeit und Umwelt (Ed.) (1999), *Beschäftigungseffekte durch Kreislaufwirtschaft*, Hannover.

Umweltbundesamt (1993), *Beschäftigungswirkungen des Umweltschutzes – Abschätzungen und Prognose bis 2000*, Texte 42/93, Berlin.

Umweltbundesamt (1997), Daten zur Umwelt – Der Zustand der Umwelt in Deutschland, Berlin.

Biomass

Andrea Marsanich

1 Introduction

Environmental impacts linked to the production, distribution and consumption of energy are definitely significant. In particular, the consumption of non renewable fossil fuels is responsible of both resource depletion, emissions and global warming. The concern for the sustainability of human activity has driven the research to the revaluation of renewable resources and less polluting energy carriers. The use of biomass for energy (electricity or heat) has been considered as a valid alternative to coal or oil, particularly with the new technological development allowing a significant reduction of emissions and a raise in efficiency of energy conversion.

Biomass energy comprises many different possible fuels derived from plant matter. Solid biomass can be burnt directly to produce electricity or heat, or it can be converted into solid, liquid or gaseous fuels using conversion technologies such as partial combustion to produce charcoal, fermentation to produce alcohol, bacterial digestion to produce biogas, and gasification to produce a natural gas substitute. Wastes from industry, agriculture and forestery can be used as a source for biomass energy, or crops such as trees and sugarcane can be grown specifically for biomass energy purposes. At present the majority of biomass energy use is in the form of the direct use of solid biomass (e.g. wood and charcoal, for heat) but an increasing proportion of biomass energy use is being directed towards larger-scale industries, such as pulp and paper and the sugar and food industries, as well as in other "modern" biomass uses such as electricity generation and the production of biofuels for transport.

Biomass accounts for about 60% of the total renewable energy contribution and about 3% of the overall energy consumption in the EU. In Italy, while the installed capacity hardly reached 20-30 MWe, in 1997, some forecasts estimate that a 2,400 MWe capacity for power generation is economically viable according to the current technologies and to the availability of biomass (Fiala et al., 72-76). The potential for heat generation and "remote district-heating" is even higher.

The resources of biomass are variable in nature and localisation. On the one hand, a relevant source of biomass is represented by wastes and by products of the wood and the food processing sectors. On the other hand, vegetal biomass can be produced on purpose for energy purposes. In both cases the indirect employment impact of the installation and the operation of a biomass fired energy generation plant can be higher than the direct one.

The purpose of the study is to test and assess a methodology for the evaluation of the impact on employment (on both the number and skill aspects) of the implementation of a biomass based energy plant. Through direct data collection, information will be gathered and analysed on the employment impact along the whole energy value chain.

In particular, the production, collection, transport and processing phases will be checked as their impact on employment may be 3 to 4 times higher than the one generated by the operations in the generation plant.

The production of energy based on biomass can be summarised by a linear value chain.

Figure 1: Production of energy based on biomass

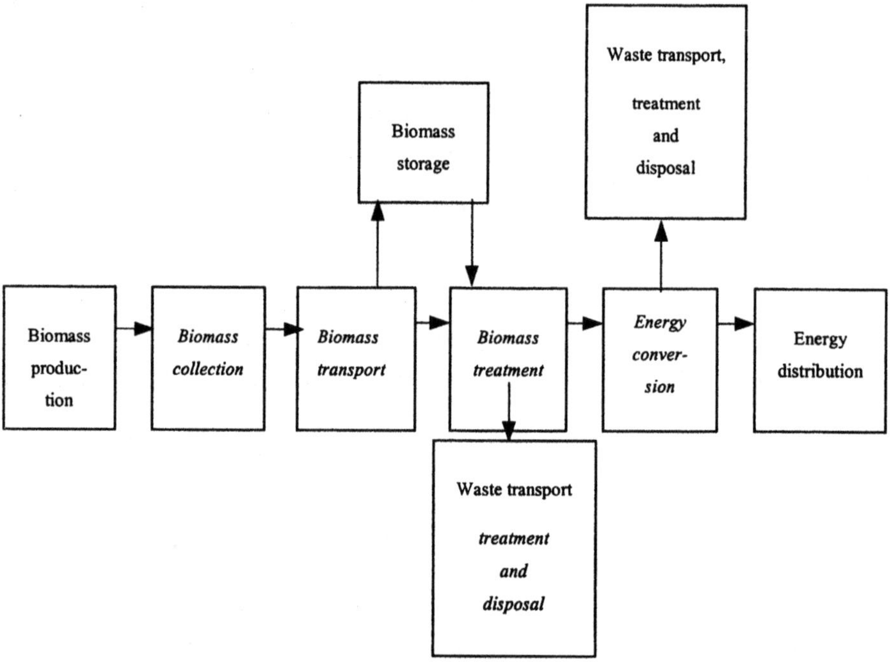

The main steps to be considered are the same that are usually analysed in the environmental impact evaluation of this technology: the production of the biomass (whether crops or residues), its collection, transport, treatment and storage, the energy conversion (combustion) and distribution.

2 Perspectives for biomass

Renewable energy is a field of great interest for the UN, EU and Italian government perspectives and policies. During the Kyoto Conference on climate change, the EU has committed itself to a reduction of 8% of Greenhouse Gas with reference to the 1990 baseline during the period 2008-2012. Therefore policies shall be developed in order to reduce the use of both energy and fossil fuels. According to the European Commission the promotion of renewable energy sources for the production of heat and electricity represents a great field of action in order to fulfil its commitment.

The European Commission's policy on renewable energy is presented in the White Book "An Energy Policy for the European Union" where it is considered as an effective instrument to achieve three main objectives: competitiveness, security of supply and environmental protection. This strategy aims at doubling the share of renewable energy.

Among the different renewable energy sources biomass can play a role within the EU and Italian action plans.

Actually a study performed by ADEME, a French energy agency, for the Institute for Prospective Technological Studies (IPTS) indicates that biomass represents about 3.5% of the total EU-15 energy consumption (about 50 Mtep), 90% of which coming from forestry, thus representing a small contribution to total primary energy demand of the EU at the beginning of the new century. Nevertheless, the study suggests that the penetration of this energy source in the market could be higher in the future if social, economic and political initiatives were jointly undertaken for its promotion. In particular, links should be drawn between the Common Agriculture Policy, energy strategy, environmental policy and employment and development policies.

The promotion of renewable energy in Italy has been developed since 1992 with the publication of the CIP 6/92, a decree providing premium prices for the electricity produced from solar, wind, biomass, or other renewable resources (Class A), from environmental assimilated energy sources as by products and wastes (Class B), or the electricity produced in co-generation (Class C). Despite those incentives the biomass based energy production, and particularly electricity production had a slow development in the last years. Actually, a certain number of plants were planned but their construction has been postponed for various reasons.

The Italian Ministry for Agriculture is performing a study that is indicating that about 250,000 hectares could be devoted to the production of biomass for energy that could represent 8 Mtep/year, directly creating 12,500 new jobs.

It is undeniable that premium prices for renewable energy could distort the market, nevertheless, once the externalities of energy production from fossil fuels have been fully internalised, the need for those incentives will be reduced.

3 The municipal heat generating plant of Dobbiaco

Dobbiaco is known as one of the coldest parts of South Tyrol. The consumption of fuel oil for heating purposes posed problems due to the high levels of pollution it induced. This was particularly due to the low efficiency of individual heating devices using diesel oil. As forestry and wood industry are well developed in the region, the municipality of Dobbiaco decided to evaluate the opportunity to implement a heat generating plant using sawmill residues as fuel.

In 1994, Tauernplan Consulting has conducted a study for a "*remote district heating system*" based on a biomass fired, heat generating plant. On the basis of the result of the study, the "*Centrale Termica di Dobbiaco sc. coop. a r.l.*" was founded. At this stage 220 pre-contracts were subscribed by the potential customers/associates. The company has been structured as a co-operative where the customers are partners and own a participation in the plant. The energy is sold for 145 Lit/kWh (0.075 €), a price based on production costs.

The construction of the heat generating plant of Dobbiaco, directed by Technisches Büro Jud, started in May 1995. The total costs for the operation amounted to 15,5 mln €, 30% of investment costs were covered by the Provincial Authority. The plant was operational in late November 1995.

The plant has been equipped with two 4 MW units using biomass fuel and one 8 MW fuel oil unit for peak demand. The technology that has been implemented in the plant is among the most advanced. The nature and quality of the biomass are continuously monitored and the combustion is automatically optimised in real time. A system of integrated filters and heat exchangers allows the achievement of thermal and environmental efficiency. All the emission levels are well below the most stringent European levels.

The heat generated is distributed to 458 households in Dobbiaco and additional 150 households are now connected to the distribution network in the nearby village of S. Candido, through a 20 km long pipeline system.

The end users' part of the system is also technologically advanced and the household has the opportunity to regulate the energy delivered through a remote control device. The system automatically regulates the energy to be produced and delivered according to the requests of the customer, thus limiting overproduction of energy. The heat delivered is directly measured and billed.

Before the construction of the plant, the houses connected to the heat distribution network usually used diesel oil and, to a lesser extent, wood for heating purposes. Fuel is still used by households but relevant economic savings have been achieved (30 to 70%).

In order to deliver an average of 14 GWh per year, the plant uses mainly by products of the wood industry, and to a lesser extent, biomass produced by local agriculture. Actually, even though the prices for the locally produced biomass are subsidised (the price is 75% higher than the market price), the amount of biomass from agriculture (28,000 m^3 per year) that was expected during the planning phase has not been achieved. Only about 1,000 m^3 have been purchased, however.

The environmental benefits linked to the adoption of such a centralised heating system have been estimated by comparing the reduction of air pollutant emissions. In particular, particulate, NOx and Carbon emissions have been reduced thanks to the higher thermal efficiency of the centralised heater (scale benefits). Moreover, the consumption of fossil fuel has been significantly reduced and the use of biomass has brought a relevant reduction in the CO_2 balance.

4 Employment impacts

According to the LCA approach, the impacts on employment, both numeric and skill effects, are considered along the "value chain". Once the boundaries have been defined, the "value chain" is divided in elementary and representative steps which are then analysed and characterised. In the present case, different steps are selected along the path going from biomass production to the final user of the heat produced, the household. Moreover, as the construction and dismantling of the plant and distribution network might represent a relevant difference compared to the reference case, those steps are considered in this evaluation.

The steps in the value chain that have been identified within the selected boundaries are summarised in Figure 2.

The reference case

In the environmental life cycle assessment, the environmental impacts are evaluated through the comparison with a reference case. In the same way, employment impacts of an innovation have to be measured in comparison with a benchmark.

As in this case, the innovation has been the adoption of a central district remote heating plant, the selected reference case will be represented by the situation before the construction of the plant (traditional single fuel heating).

The employment impact has been assessed including the construction, and maintenance phase of the plant and of the pipeline network. In the same way, when dealing with the employment impact, one should consider the alternative use of the biomass, both the one produced by agriculture and the one obtained from sawmills. In particular, the sawmill by products were formerly used by the pulp and paper industry and thus the effect in this sector has been addressed.

Figure 2: Value chain

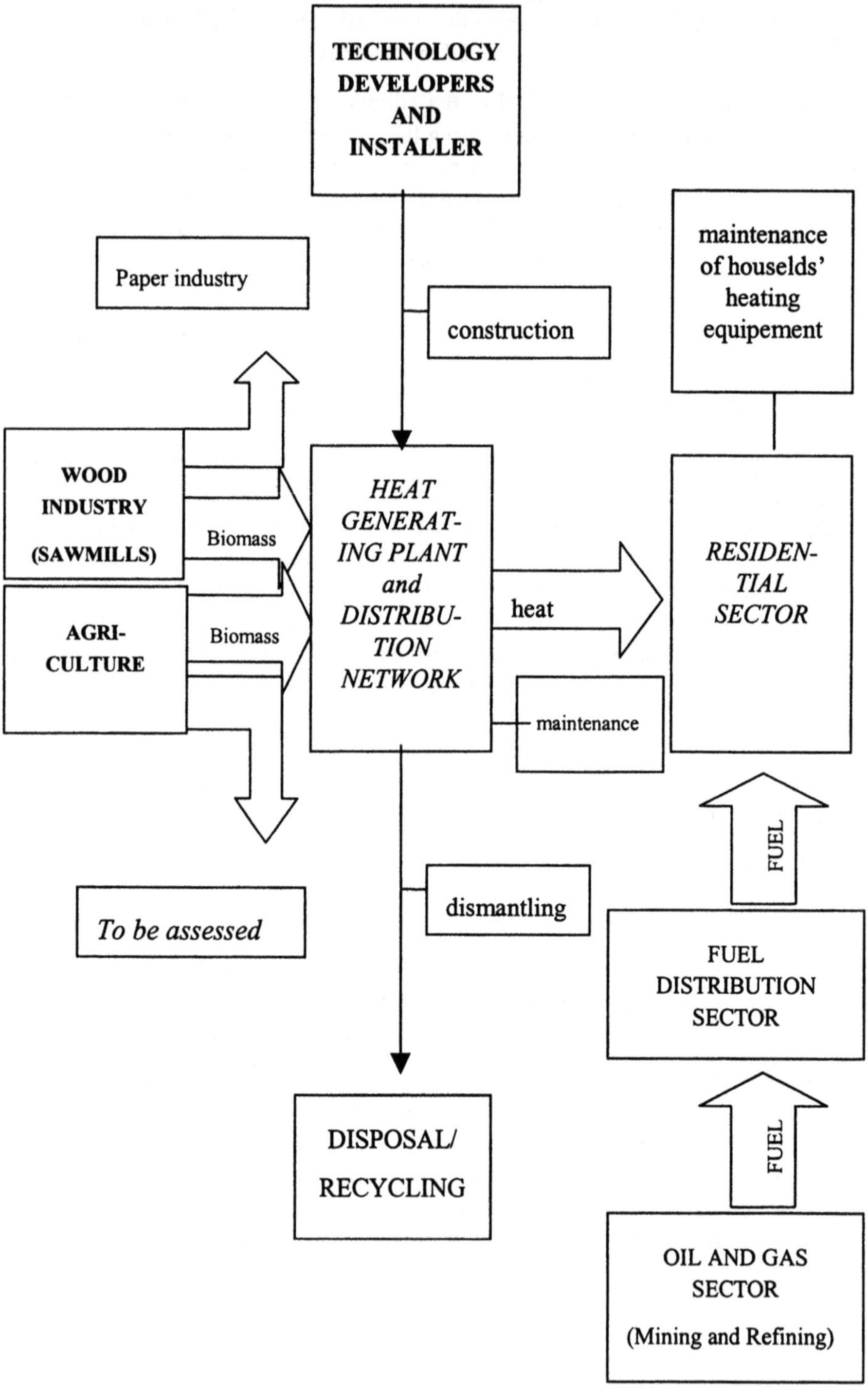

Moreover, the impact of the reduced fuel consumption (2.5 mln l of diesel oil) due to changes in heating devices has to be considered both with regard to the maintenance of those devices and to the production and distribution of diesel oil.

4.1 Direct impacts

4.1.1 The production of biomass in sawmills

The biomass used in the plant comes mainly from sawmills (in Dobbiaco, S. Candido and in the nearest part of Austria) as wooden residues. The plant only accepts sawdust and small chips. Before the construction of the plant, the wooden residues were sent to pulp and paper factories in Longarone and other locations (>80 km). The price paid by the heating plant is the same that is paid by the pulp and paper industry, but, as it includes transport it brings more benefits to the sawmills as the distance to be covered is significantly shorter (<40 km).

Numeric impact

No direct employment impact (positive or negative) has been observed or has been considered to be due to the operations of the biomass plant during the interviews. Nevertheless, the implementation of the plant has an economic impact on the sawmills as an increase in income has been observed. In the medium/long term, these economic impacts could bring some income related employment effects which are hard to evaluate at his stage.

Skill impacts

No skill impact has been observed as the biomass sold to the heating plant does not require different treatment or processing than the one that was previously sold to the pulp and paper industry.

4.1.2 The production of biomass in agriculture

The Dobbiaco Heating Plant buys agricultural residues (particularly from forestry) at a price which is 75% higher than actual market prices. Despite these premium prices this source of biomass plays a limited role in the supply of the plant. One particular reason is linked to the fact that the biomass must be brought to the plant in small chips or dust, therefore requiring additional processing the suppliers usually are not able to perform, both for cost and skill/equipment reasons.

Numeric impact

As agriculture plays only a limited role in the supply chain of the plant, no significant numeric impact has been observed.

Skill impacts

In the same way, no skill impact has been observed in agriculture.

4.1.3 Transportation of biomass

Before the construction of the heating plant in Dobbiaco, the wooden residues were brought by lorry to Longarone or other pulp and paper factories in the region. The average distance covered for this transport was about 80 to 130 km. As already mentioned, the opportunity to sell the by product in the surroundings allowed a significant reduction in costs for sawmills. On the other hand it has undoubtfully reduced the labour intensity of biomass transportation. In particular the time spent for transportation has been reduced by 3 to 6 times.

Numeric impact

The net effect of the change in final destination of biomass is not precisely quantified. In particular, the majority of suppliers use internal human resources for the transportation of residues to the plant or to paper industry. In terms of job losses, there has been no real impact as no reduction has been observed. Nevertheless, in terms of labour intensity, the reduction of the transport distance has significantly reduced the amount of time spent for biomass transportation to its final destination.

Skill impacts

As the only changes observed are linked to the final destination of by products, no skill effect has been observed.

4.1.4 Treatment of biomass

The biomass that is used in the heating plant of Dobbiaco does not require particular processing as it is mainly composed by dry wooden residues and sawdust. Moreover, the plant has designed such that it is able to immediately burn the residues from sawmills and wood industry in the form they were previously delivered to the paper industry. Thus, the impact on employment of this stage of the "biomass value chain" is not significant.

Numeric impact

No significant employment effects are linked to this stage of the value chain.

Skill impacts

No significant skill impacts are observed at this stage.

4.1.5 Energy production

- The plant of Dobbiaco uses advanced technologies and has a high level of automation. Despite the number of customers it supplies and the amount of energy it delivers, it only has 4 full-time employees. The operation of the plant and the network, and its ordinary maintenance are directly performed by those four employees.

Numeric impact

- At the start-up of the plant, three persons were appointed. During the first year, a fourth person has been employed as the number of households connected to the network has been raising. The total employment directly generated within the plant reaches 7,480 man-hours.

Skill impacts

The employees have received training courses, because electronic devices and software are used in the plant.

4.1.6 Energy distribution

The distribution of energy is automatically managed by the plant and the remote control by the single households. No additional employment is required for the distribution of the energy.

Numeric impact

The numeric impact of this step of the value chain is included in the energy production part.

Skill impacts

No skill impact can be observed in this stage.

4.1.7 Construction

The construction of the plant started in May 1995. In April 1996, 70% of the plant and network were operational. The company has provided an estimate of the labour required for the construction of the pipeline network and of the plant itself: about 24,000 man-hours for the network, and approximately 11,000 for the plant. The labour required for the construction of the equipment has not been accounted for as it was considered beyond the boundaries of the analysis.

Numeric impact

The labour required by the construction of the plant and the pipeline network has been estimated to reach 35,000 man-hours. This employment has been transitory, nevertheless it can be calculated if the lifetime of the plant is supposed to be about

10 or 15 years. Moreover, the evaluation of the labour required for the construction is used to have an estimate of the maintenance requirement during the lifetime of the plant.

Skill impacts

No particular skill impact is linked to the construction of the plant.

4.1.8 Maintenance

The normal maintenance of the plant and the pipe network is performed by the full-time employees of the plant. Nevertheless, extraordinary maintenance during the lifetime of the plant has been considered. In particular, an estimate has been taken at about 3,500/4,000 man-hours in ten years. This represents only a rough estimate as no particular accident occurred during the first years of operations. More precise data shall be provided by the building company.

Numeric impact

A rough estimate indicates a 3,500/4,000 man-hours in ten years.

Skill impacts

No particular skill impact is attributable to maintenance activities.

4.2 Other impacts

In order to have a complete and comprehensive description of employment effects of the impact of the adoption of this new technology in households' heating we must consider the use of biomass now used by the plant, as well as the alternative fuels that were previously used in the region for heating purposes. Before the construction of the plant, by products of sawmills were used mainly by the pulp and paper industry. Therefore factories have been contacted in order to get information on the effect brought by the reduction of the availability of biomass in the region.

In the same way, the two main diesel oil suppliers of the region have been interviewed in order to estimate the scale of the negative effect linked to the reduction in sales.

4.2.1 Indirect effects in the oil and gas value chain

Before the start of the plant, the inhabitants of Dobbiaco used individual diesel oil fired heaters. The oil consumed for heating purposes has been significantly reduced. The implementation of remote heating based on biomass has allowed a reduction in diesel oil consumption of about 2.5 mln l.

In order to evaluate the dimension of this negative effect on the oil distribution sector, the two major suppliers of the region have been contacted. The reduction of diesel oil demand has reduced their income to a level that is now threatening the employment of drivers. In particular, as the diesel oil was previously sold to single households, the reduction of 2.5 mln l represents, according to the contacted companies about 1,500 to 3,000 man-hours per year. Those evaluations have been calculated by the companies according to the number of trucks that previously were sent to Dobbiaco and to the average time needed for this transport activity.

No jobs have yet been lost, nevertheless the representative of one of the companies expects that in the near future they could have to reduce the number of drivers they employ.

Numeric impact

A net negative effect has been observed in this sector. The losses have been estimated at somewhere between 1,500 to 3,000 man-hours per year.

Skill impact

No skill effects have been observed.

4.2.2 Indirect impacts in the Pulp and paper industry

The pulp and paper industry was the main client for sawmills wooden residues before the implementation of the heating plant of Dobbiaco. The fact that the supply from Dobbiaco has been reduced has not affected the employment or the income of this sector. Actually, in the region there is a relevant availability of biomass thus the paper mills have a significant number of alternative supply channels. Therefore no real impact can be observed in this sector.

Numeric impacts

No significant employment effects have been observed in this sector.

Skill impacts

No significant employment effects have been observed in this sector.

4.2.3 Indirect impact in other sectors

As the use of single households' heating devices can be reduced by the development of a centralised remote heating network, a certain effect can appear in their maintenance and replacement. This can lead to a reduction on the employment side in the heating device production, distribution and maintenance industry. While the two first aspects are hard to assess, the reduction of labour requirement for the maintenance of diesel oil fired individual heating devices has been calculated through direct interviews with operators of the sector. The maintenance required for the 600 households has been reduced at least by 30/40%. Considering that, in such a cold region a diesel oil heating device requires from 5 to 10 hours a year for maintenance, the reduction has to be estimated at between 900 and 2,400 man-hours per year.

5 Conclusion

The implementation of the Heat Generating Plant of Dobbiaco has brought undeniable positive environmental effects, at both the local (improvement of air quality) and global level (enhancement of the CO_2 balance). In the same way it has brought a certain level of economically positive effects as it has reduced the dependence on foreign energy supply.

In the same time the plant has brought a set of employment effects, positive and negative. Table 1 reports the employment effects observed or evaluated for each step of the selected value chain.

Table 1: Employment effects of value chain steps

Step	Numeric impact	Skill impact
Production of biomass in sawmills	=	=
Production of biomass in agriculture	=	=
Transportation of biomass	=/– (non quantified)	=
Energy production	+ (7,480 man-hours/year)	+ (IT training courses)
Energy distribution	=	=
Construction	+ (35,000 man-hours in 10 years. transitory – to be quantified)	=
Maintenance	+ (3,500-4,000 man-hours in 10 years)	=
Diesel oil distribution sector	– (1,500-3,000 man-hours per year)	=
Pulp and paper sector	=	=
Households' heating device maintenance	– (900-2,400 man-hours per year)	=
Other sectors	+ (due to income effects – not quantifiable)	
= no relevant impact; – negative impact; + positive impact		

Considering that part of the employment effect is limited to the construction phase, a balance could be made in a ten years timeframe – which represents the minimum lifetime of the generating plant.

The positive effects in ten years, could be: direct employment within the plant (4 full-time employees*10 years ≈. 74,800 man-hours) + labour required in the construction phase (≈ 35,000 man-hours) + labour required by the extraordinary maintenance (≈ 3,500 man-hours).

The negative effects are: reduced labour requirement for the transport of diesel oil (≈ 15,000/30,000 man-hours) + reduced labour requirement for the maintenance of heating devices (≈ 9,000 to 24,000 man-hours).

The final balance seems to indicate that a net positive effect can be observed in the lifetime of the plant, when the construction phase is included.

Finally, one must consider the fact that a plant like the one implemented by the municipality of Dobbiaco can bring a positive income effect in the region. In particular, the plant has enhanced the energy autonomy of Dobbiaco bringing a certain level of increase for the local income. Moreover, the lower expenses for energy have enhanced households' revenues leading to a potential increase in the demand for other goods or services.

Reference

Fiala, M., G. Lai, G. Pellizzi and G. Riva (1995), Biomasse per la produzione di energia elettrica, *Energia* 3/95, 72-76

Biotechnology

Anthony Arundel, Ivo Demandt and René Kemp

1 Introduction

This chapter offers an overview of the employment effects of four types of biotechnology for industrial production and agriculture. Employment effects and environmental effects are assessed at varying points of the value chain (from extraction, cultivation, processing, use, and waste management after use) but the focus is on employment effects at the point of production of the biotechnology and during its use.

We departed from the original goal of the case study on biotechnology, which was to focus on one type of biotechnology for industrial use. Two biotechnological applications had originally been considered for closer analysis: bio-bleaching in the pulp and paper sector and the use of improved plant crop varieties in the starch industry. Unfortunately, it was not possible to meet this goal for two reasons. First, although clean industrial process biotechnology has received extensive publicity, the reality is that many of these clean technologies are in the pilot phase and have not yet been applied on a wide scale[1]. Second, several firms involved in the development of genetically-modified crops for starch refused interviews because they did not wish to attract attention, given the current controversy in Europe over agro-biotechnology. For both reasons, we decided not to conduct an in-depth case-study of one clean production biotechnology that would carefully follow employment effects throughout the value-added chain. As an alternative, we take a closer look at direct and indirect employment effects at the point of use and the point of production for a wider range of biotechnologies with environmental benefits. This means that we have four case studies instead of one, allowing for a broader assessment of the employment effects of biotechnology. The industries studied are: pulp and paper, industrial starches, and fine chemicals. The fourth study is on agricultural biotechnology.

The four case studies are based on a review of the relevant literature, interviews with firm managers, and, in the case of agricultural biotechnology, analyses of a

---

1 See, for example, the discussion of several biotechnology applications to clean production in the OECD report *Biotechnology for Clean Industrial Products and Processes* (OECD, 1998). The reality of low biotechnology use in the paper industry is discussed by Laestadius (2000).

survey of seed firms and European field release data for genetically-modified (GM) plants.

Biotechnology involves the use of biological organisms, systems and processes to facilitate industrial, pharmaceutical, and agricultural processes. Biotechnological processes offer a range of environmental benefits, both through end-of-pipe applications to clean polluted soil, water or air and in clean production technologies. An example of the latter is the use of enzymes in industrial and food processing. Environmental benefits can occur through the use of less environmentally harmful feedstocks, lower temperature operations which can save energy, and through improved recycling.

It is important to have a good definition of 'biotechnology'. Sharp (1991) discusses three different 'biotechnologies'. In industrial applications, the first generation consists of simple processes that have been in use for several millennia to make beer and cheese, while second generation biotechnologies include more complex systems based on products produced by micro-organisms, such as the use of enzymes in manufacturing. The third generation is generally assumed to be based on genetic engineering, although other technologies such as peptide synthesis are usually included. Often, first, second and third generation biotechnologies can be used to achieve the same result, creating alternative technological choices.

The use of biotechnology in health applications has attracted the lion's share of biotechnology investment in Europe and North America (Muller et al., 1997, Morrion and Giovanetti, 1998). Yet the future environmental and employment impacts of advanced biotechnology is probably greatest in several resource-based sectors, which include both extraction industries such as mining and forestry and resource-based manufacturing sectors such as petroleum refining and pulp and paper (Arundel and Rose, 1998, Arundel and Rose, 1999, Autio et al., 1997, CBS Taskforce, 1997, Laestadius, 2000, Tils and Sorup, 1997), and in the agro-food sector (Burke and Thomas, 1997). The potential environmental benefits of biotechnology for industry are due to better end-of-pipe and clean production technologies. The environmental benefits of agricultural biotechnology can occur throughout the agro-food chain and in industry, as when crops replace industrial feedstocks.

Biotechnological innovation essentially replaces a chemical, mechanical, or agricultural process with a biological process. This means that most biotechnology innovations are unlikely to be adopted unless they can offer superior quality or cost savings in comparison with existing processes, although there are a few exceptions (Arundel, 2000). Genetic engineering can create completely new drugs while some applications of agricultural biotechnology can create plant varieties with new uses. Nevertheless, most biotechnological innovations are likely to save labour at some point in the value-added chain.

As noted, this chapter examines the employment effects of three types of industrial biotechnology use; in pulp and paper, industrial starches, and fine chemicals, plus agro-biotechnology. The first three cases all involve the use of enzymes which can be produced by 'wild' strains of bacteria or by genetically-engineered bacteria. The fourth case involves the use of genetic engineering and associated techniques to develop new crop varieties that either could not be developed using conventional breeding or which would take several years longer. A short explanation of enzyme technology is provided below before proceeding to the case studies.

2 Biotechnology of enzymes

Enzymes are proteins that consist of long chains of amino acids held together by peptide bonds. They are present in all living cells, where they control the metabolic processes whereby nutrients are converted into energy and new materials. Furthermore, enzymes take part in the breakdown of food materials into simpler compounds. Some of the best-known enzymes are those found in the digestive tract where pepsin, trypsin and peptidases break down proteins into amino acids, lipases split fats into glycerol and fatty acids, and amylases break down starch into simple sugars.

Enzymes are capable of performing these tasks because, unlike food proteins such as casein, egg albumin, gelatine or soya protein, *they are catalysts*. This means that by their mere presence, and without being consumed in the process, enzymes can speed up chemical processes that would otherwise run very slowly, if at all. After the reaction is complete, the enzyme is released again, ready to start another reaction. In principle, this could go on forever, but in practice most catalysts have a limited lifetime. Sooner or later their activity becomes so low that it is no longer practical to use them. This is particularly true for industrial enzymes. Most are therefore used only once and discarded after they have done their job.

Contrary to inorganic catalysts such as acids, bases, metals and metal oxides, enzymes are very specific. In other words, each enzyme can break down or synthesize one particular compound. In some cases, their action is limited to a specific chemical bond. Most proteases, for instance, can break down several types of protein, but in each protein molecule only certain bonds will be cleaved depending on which enzyme is used. In industrial processes, the specific action of enzymes allows high yields to be obtained with a minimum of unwanted by-products.

Enzymes are part of a sustainable environment, as they come from natural systems, and when they are degraded the amino acids of which they are made can be readily absorbed back into nature. Fruit, cereals, milk, fats, meat, cotton, leather and wood are some typical candidates for enzymatic conversion in industry. Both the usable products and the waste of most enzymatic reactions are non-toxic and readily broken down. Finally, industrial enzymes can be produced in an ecologically sound way where the waste sludge is recycled as fertilizer.

A major environmental advantage of enzymes is that their catalytic properties occur at comparatively low temperatures, between 30-70°C, and at pH values that are near the neutral point (pH 7). For certain technical applications, special enzymes have been developed that work at higher temperatures, although no enzyme can withstand temperatures above 100°C for long. These characteristics mean that processes based on enzymes can result in energy-savings and lower capital equipment costs, since reactors do not need to be resistant to heat, pressure or corrosion.

One disadvantage of enzymes for environmental applications is that they do not work well under cool conditions. This limits their use in cold climates such as in northern Europe for resource extraction such as mining (CBS, 1997).

2.1 Research and development

New techniques such as genetic engineering and the related discipline of protein engineering are speeding up the product development cycle for new enzymes. Enzyme research specializes both in new techniques of molecular biology as well as the classical ones such as the screening of micro-organisms.

When a new enzyme or enzyme application has been discovered, it has to be evaluated under practical conditions. Upscaling from small batch conditions to large-scale use is therefore a vital developmental step. Industrial processes may need to be optimized for the use of enzymes. The selection of the right enzyme and the establishment of optimum process conditions are of great importance. Another area of importance is the formulation and granulation of enzyme products. Enzymes have to be stabilized so that the finished product can be shipped and stored without loss of enzymatic activity.

2.2 Enzyme production

The starting point for production is a vial of a selected strain of microscopic organisms that produces a desired enzyme. They will be nurtured and fed until they multiply many thousand times. After fermentation, the enzyme is separated from the production strain, purified and mixed with inert diluents for stabilisation. Then the desired end-product is sold as a standardised product.

Many types of enzymes are produced by GM micro-organisms. These enzymes are produced under well-controlled conditions in closed fermentation tanks. Due to the efficient purification process in which the enzyme is separated from the production strain, the final product does not contain any GM material.

It is in R&D and the production of enzymes that we should expect the most significant employment effects.

2.3 Environmental benefits of enzymes

Enzymes offer four potential environmental benefits:

- Enzymes work best at mild temperatures and under mild conditions. They can be used to replace high temperature conditions and toxic chemicals, thus saving energy and preventing pollution.
- Enzymes are highly specific, which means fewer unwanted side-effects and by-products in the production process.
- Enzymes can be used to treat waste consisting of biological material.
- Enzymes themselves are biodegradable, so they are readily absorbed back into nature.

2.4 Industrial applications of enzymes

Enzymes have a wide range of industrial applications in detergents, textiles, starches and sugar, food and feed, pulp and paper, leather, health care products, and fine chemicals. The next three sections provide case studies of the employment effects of enzymes used in pulp and paper, starches, and fine chemicals.

3 Pulp and paper

3.1 Introduction

The most commonly used raw material to produce pulp is wood, which mainly consists of cellulose, hemicellulose and lignin. Wood fibres contain cellulose and hemicellulose. Lignin can be thought of as the glue holding the wood fibres together. Another component is pitch, which acts as a tree's defence mechanism against microbial attack.

In the pulping process the wood fibres are brought into suspension – the pulp. There are two different types of pulping processes that can be used. First there is mechanical pulping which separates the fibres mechanically with the input of large amounts of energy. Mechanical pulps are often called high-yield pulps since all the wood components are conserved in the pulp, including the lignin. They are less expensive to produce than chemical pulps, but they have the disadvantage that they become darker when exposed to sunlight. They are used mainly in the manufacture of newsprint and magazine paper. Second there is chemical pulping in which wood chips are cooked in chemicals until the lignin dissolves, releasing the wood fibres. The dominant chemical pulping process is the kraft process, which gives a dark brown pulp due to the residual lignin. This residual lignin must undergo some type of bleaching process to yield a bright, white wood pulp before it can be used for fine paper manufacture. In one end-use, it will be converted into fine paper grades (Sappi, personal communication, Novo Nordisk).

A recent survey by Statistics Canada shows that the pulp and paper sector is a major user of biotechnology. However, most use is confined to environmental biotechnology[2], particularly for pollution control. Only 13% of employees worked in a firm that used biotechnology in an industrial process (Arundel and Rose, 1999). Many of the latter applications of biotechnology involve the use of enzymes in the manufacturing process. These enzyme-based systems are the focus of this section.

3.2 Enzyme applications in pulp and paper

Until recently, the use of enzymes in the pulp and paper industry was not considered technically or financially viable. Except for the limited use of enzymes to modify starch for paper coatings, suitable enzymes were not readily available. However, driven by market demand and environmental standards, new enzymes could offer significant benefits for the industry. Possible applications involving enzymes are biopulping, enzymatic pitch control, enzymatic deinking of waste paper, bleach boosting, and improving paper strength and drainage rates.

3.2.1 Biopulping

Several processes are used to separate the cellulosic fibres from the lignin in wood to form a slurry that is further processed into paper. The existing chemical processes are particularly polluting. In biopulping lignocellulosic materials are being treated with lignin-degrading fungi to manufacture the pulp. This fungal treatment could result in energy-savings and improved paper strength and is clearly a cleaner process as it saves on chemicals.

The economic feasibility of biopulping has been demonstrated at a pilot scale. The process increases the mill throughput by 30% or reduces the electrical energy requirement by at least 30% at unchanged output (OECD, 1998).

The use of biopulping potentially could lead to some reduction in employment upstream in the production of chemicals, which then would be compensated for in the development of enzymes. Also the increase in energy efficiency might lead to a lower demand for energy lowering employment in the upstream energy sector. However, the increased energy efficiency in pulping could also be used to increase

---

2 The environmental biotechnologies used by pulp and paper respondents consist of bio-augmentation, bio-reactors, bio-remediation, and phyto-remediation.

output. In this case the effect on employment in the energy sector would be neutral.

However, the driver for a switch to biopulping will clearly not be its possible effect on employment or its positive effect on product quality. Instead it could be driven by stricter environmental legislation on the use of chemicals, pressure to save on energy reducing CO_2 emissions, and a desire to reduce production costs. Few employment effects are expected from biopulping within the pulp and paper industry.

3.2.2 Enzymatic pitch control

Pitch is a mixture of hydrophobic resinous materials found in many wood species, which cause a number of problems in pulp and paper manufacture. Pitch agglomerates form on the processing equipment such as the chests, felts and rollers. These agglomerates can cause holes in the paper so it has to be recycled or downgraded in quality. In the worst cases, the paper web can break, causing costly paper machine downtime.

Traditional methods of controlling pitch problems include natural seasoning of wood before pulping and/or adsorption and dispersion of the pitch particles with chemicals in the pulping and paper making processes, accompanied by adding fine talc, dispersants and other kinds of chemicals (RPE, personal communication, OECD, 1998). During the past ten years or so, biotechnological methods have been developed and are now being used industrially. A commercial lipase has been developed for use in mill operations. This enzyme has proved its ability to reduce pitch deposits significantly on rollers and other equipment. It breaks down triglycerides in wood resin in the pulp in much the same way as fungal and bacterial growth reduces the pitch content of the wood during conventional seasoning. However, unlike seasoning, where the wood is stored for a long time, the enzyme acts immediately and does not reduce brightness or yield. In the early 1990s, Sandoz introduced a new product which metabolises pitch quite effectively by lignin-degrading fungi in biopulping, thus offering an additional benefit (Novo Nordisk, OECD, 1998).

Enzymatic pitch control replaces the use of chemicals by enzymes to reduce wood pitch. As such there might be a substitution of labour from chemical production toward enzyme production. As enzymatic pitch control would make the seasoning of wood superfluous, the process of storing wood to reduce pitch becomes redundant, which could lead to reductions in employment in that area. However, the industry is not likely to switch to enzymatic pitch control to save on labour costs, since such savings are very small. A reduction in operational problems and possible restrictions on the use of chemicals could be stronger motivations to use enzymes to control pitch.

3.2.3 Enzymatic deinking

Enzymes have a large potential for the deinking of waste paper. Traditional deinking uses caustic soda, silicates and peroxide for oil-based printing materials such as newspapers and magazines. With the growing use of coatings and new types of inks containing synthetic polymers, conventional deinking is inadequate for producing high-quality pulps. Moving to a enzymatic deinking which can employ neutral or alkaline enzymes requires some change in the chemistry of the system, but can result in improvements in both the process and the final product. This can include improved pulp cleanliness, improved operation of the grey-water loops, less deposit potential and a brighter final pulp (Novo Nordisk, OECD, 1998).

Again a possible employment effect could be a substitution of labour from chemical production to enzyme production. The employment effect within the pulp and paper industry is likely to be negligible, even though it involves an extra process step, because of the high degree of automation and computerisation within the industry. Stricter regulation on the use of chemicals and eventual limits to traditional technologies could drive firms toward applying enzymatic deinking.

In fact the need to deink can in many cases be avoided. Paper manufacturers producing high-quality paper can use virgin fibres, while those using recycled material serve different markets, such as packaging, which do not require the same product standards as for fine papers.

3.2.4 Bleach boosting of kraft pulps

Kraft pulps account for most of the world's pulp production. Their characteristic brown colour must be removed by bleaching before the manufacture of fine papers. Chlorine and derivatives of chlorine have been the cheapest and most versatile bleaching agents available for the bleaching of chemical pulps. This class of compounds has the disadvantage of forming chlorinated organic substances (some of which are toxic) during bleaching. Due to consumer resistance and environmental regulation on chlorine bleaching, pulp-makers are turning to other bleaching processes, such as elemental chlorine free or totally chlorine free bleaching, to extended pulping times, and to other process modifications. The disadvantages associated with some of these methods are higher costs and/or greater loss of pulp yield and strength as compared with chlorination. (OECD, 1998, TNO, personal communication).

By treating the kraft pulp enzymatically (mainly xylanases) prior to bleaching, it is possible to obtain a very selective partial hydrolysis of the hemicellulose, which has precipitated onto the fibres during the kraft cooking process. The enzyme has two indirect effects – first, it is possible to wash out more lignin from the pulp, and, second, the pulp becomes more susceptible to bleaching chemicals. The technique is called 'bleach boosting' and gives a significant reduction in the need for chemicals in the subsequent bleaching stage, with almost no loss in pulp yield or quality (Novo Nordisk, OECD, 1998).

Bleach boosting is a clear case in which restrictions on the use of traditional chlorine chemicals have led pulp and paper manufacturers to look for alternative processes. Still, firms will first consider chemical-based alternatives such as elemental chlorine free or totally chlorine free processes. Whether or not enzymes will replace chemical processes will depend on the costs and performance of enzymes compared to chemicals. One estimate is that enzymes reduce total pulping costs by between 1% and 3%. This slight benefit could be insufficient to overcome resistance to biotechnology, based on a lack of experience, in many pulp and paper firms (Laestadius, 2000). Tougher environmental legislation could be necessary to give enzymes a larger cost advantage.

Again the employment effects consist of upstream effects due to a substitution of labour from chemical toward enzyme production. There will probably be no perceptible effect at all within the pulp and paper industry.

3.2.5 Improving paper strength and drainage rates

The structure and chemical composition of pulp fibres are very important for paper strength and other properties. Enzymes can be used to improve physical properties of fibres and might have a commercial role in the future. For example, cellulases and xylanase can enhance pulp fibrillation and thereby improve paper strength. They can reduce fibre coarseness and increase paper density and smoothness. Starch-modifying enzymes are sometimes also used to improve paper quality. These applications could lead to increased employment in the upstream enzyme producing industries.

The speed of paper machine operation depends in part on the drainage of water out of the pulp mat. Treating cellulose fibres with cellulases and hemicellulases allows water to drain more quickly from the wet pulp, thereby reducing processing time and energy use for drying (OECD, 1998).

As for biopulping, improving drainage rates could lead to reduced employment in the upstream energy sector. There will probably be no effect on employment in the paper and pulp industry.

3.2.6 Starch modification for paper coating applications

In the manufacture of coated papers, a starch-based coating formulation is used to coat the surface of the paper. The coating provides improved gloss, smoothness and printing properties compared to the uncoated product. Raw starch is unsuitable for this application, since the flow properties would be unsuitable. In one case, chemically modified starch with a much lower solution viscosity is used. As an economical alternative to modifying the starch with aggressive oxidizing agents, the starch can be treated with enzymes (alpha-amylases) to obtain the same viscosity reduction (Novo Nordisk).

Chemical modification of starch can either happen at the starch producers or at the paper mill using a batch or continuous process. For starch to react with enzymes it has to be cooked first. The cooking of starch is an integral part in the paper-making process, whereas for starch producers it is quite inconvenient as it would involve a couple of extra process steps. Therefore enzymatic modification normally would have to take place at the paper mill (Cargill and Cerestar, personal communication).

Whereas chemical modification is more harmful to the environment as it uses chemicals that have to be washed out of the effluent in a later stage, enzymatic modification needs an extra process step to stop the process as enzymes are self-propagating (RPE, personal communication).

Both types of modification reduce the BOD of the effluent as they improve the attachment of starches to the wood fibres.

The employment effect is limited to some upstream substitution of labour between chemical production and enzyme production.

3.2.7 Other applications

There are interesting possibilities for future applications of enzymes in the pulp and paper industry. One possibility is the selective action of an endo-cellulase, which can improve individual fibre characteristics, for example, in producing a softer tissue product. Furthermore, other types of carbohydrate are reported to reduce the amount of energy required for pulp refining, or in reducing unwanted components such as vessel segments, which can cause printing problems with the final paper.

3.3 The impact of enzymes on employment

According to the literature and to several of the interviews, the application of enzymes in the pulp and paper industry could lead to a broad range of benefits. The introduction of biopulping, bleach boosting and enzymatic deinking could significantly reduce the need for chemicals. Biopulping and enzymes to reduce drainage rates could lead to quite substantial energy-savings. Other potential benefits of using enzymes mainly involve improving paper quality.

The employment effects of these applications within the industry are expected to be minor and insignificant, as shown in Table 1, and almost entirely limited to upstream effects. These include employment losses in the energy sector due to the energy-saving potential of some enzyme applications. Others concern substitution effects between enzyme and chemical production due to the potential of some enzyme applications to save or replace chemical use. There may also be a substitution effect in waste management: the use of enzymes leads to less chemical waste in need of treatment but an increase in organic matter that needs to be taken care of.

All these applications, however, are still in an experimental stage of development. The firms interviewed did not use them at this moment, although they were seriously considering some of them. Therefore the effects on costs, employment and environment mentioned previously in this section are mainly speculative and preliminary. In the future, learning curve effects could give some enzyme applications a competitive edge and facilitate their widespread adoption. No major cost reductions are foreseen at this moment though.

A lack of significant employment effects is supported by the results of the Statistics Canada survey, which found that only 14% of employee-weighted wood, pulp and paper firms reported cost savings from the use of environmental biotechnology, compared to 87% of mining firms, 69% of petroleum, gas and coal firms, and 61% of food firms (Arundel and Rose, 1999).

Positive employment effects will probably be concentrated in the R&D stage of enzymes, which takes place at biotechnology firms upstream and not within the industries themselves. Due to the high degree of automation and computerisation in the pulp and paper industry, switching from chemical to enzymatic processes will not have any significant impact on employment. Despite the fact that biotechnology involves quite advanced technologies it also has no perceptible effect on the skill level of the labour force. All this may change when biotechnology will achieve a higher grade of penetration and gain in importance in the pulp and paper industry. Only then the industry may have to internalise R&D and the expertise with regard to biotechnology, leading to increased employment. As for now user industries can simply buy the processes they need from biotechnology firms, like Genencor, Gist Brocades and Novo Nordisk.

The success of enzyme applications will ultimately depend on their costs compared to their traditional chemical alternatives. Only enzymes that are produced on a large scale can in fact effectively compete with chemical alternatives. Unfortunately they are relatively few in the pulp and paper industry. Consequently, the industry will in most cases prefer chemicals over enzymes, unless there are severe environmental restrictions on the use of these chemicals increasing the costs of their application.

Furthermore, in the Netherlands the potential of the application of enzymes is limited to those that involve the paper making process, because the pulp to produce different kinds of paper and board is imported from elsewhere. The only process in the Netherlands in which enzymes are currently considered is in the modification of starches to improve its capability to bind wood fibres.

Table 1: Overview of employment effects related to the use of enzymes in the paper and pulp chain

	Biopulping	**Enzymatic pitch control**	**Enzymatic deinking**	**Enzymatic bleach boosting**	**Improve paper strength and drainage rates**
Natural resource and extracting industries	-- in forestry: less wood needed due to higher yield	0	0	– in foresty: less wood needed because of higher yield	0
Intermediate goods industries	+ enzyme production – chemical production – reduced energy use	+ enzyme production – chemical production	0	+ enzyme production – chemical production	+ enzyme production – chemical production – reduced energy use
Paper and pulp industry	0	0	0	0	0
Waste processing	– less treatment of chemi-cal waste	– less treatment of chemical waste	– less treatment of chemical waste	– less treatment of chemical waste	– less treatment of chemical waste
Compensatory effects	0	0	0	0	0
Overall effect	–	0	0	–	0
Main driver	Environmental concerns	Operational concerns	Product and process improvements	Environmental concerns	Energy saving and reduced process time

0 no effect on employment, – small reduction in employment (< 2%), + small increase in the level of employment (< 2%) -- reduction in employment between 2-5% ---large reduction in employment (> 5%).

4 Industrial starches

4.1 Introduction

The raw materials for the extraction of starch are maize and wheat, but it is also possible to use potatoes. Maize is the ideal raw material for starch extraction and is used in the US, but more wheat than maize is grown in Europe (the value of the wheat crop is approximately 2.5 times greater than the maize crop), due to climatic differences and the effect of agricultural subsidies (OECD, 1999, DG Agriculture, 2000a). Whereas starch can be extracted from maize mechanically, it is necessary to use enzymes to achieve the same yield in extracting starch from wheat. Cellulases are used to improve the yield of starch extraction from wheat. Without the possibility of using enzymes, the extraction of starch from wheat would not be financially feasible.

Next to maize and wheat, potatoes also can be used for starch extraction. This route has been pioneered by AVEBE, a Dutch company, probably due to the availability of potatoes in the Netherlands. Although it is more expensive to use potato starch, it has quite favourable characteristics. As such potato starch seems to be more amenable to enzymatic modification. Furthermore AVEBE has developed a GM potato for use in the starch industry. Unfortunately, the commercialisation of this potato has been delayed since the 1999 *de facto* moratorium in the European Union on the commercial use of GM crops.

The extracted starch is either converted into different kinds of syrup or it is modified or simply sold in its native form for use in the pulp and paper industry and the food industry. Whereas the modification of starches for the pulp and paper and the food industry currently is primarily chemical, starch conversion to produce syrups is nowadays mainly enzymatic.

4.2 The history of starch conversion

As early as the beginning of the 19th century, it was discovered that by boiling starch with acid it could be converted into a sweet-tasting substance, which consisted mainly of glucose. This product, however, did not provide a complete substitution for sugar, partly because glucose is only about two-thirds as sweet as cane or beet sugar and partly because the yield using his technique was not very high.

Nevertheless, since then acids have been used widely for breaking down starch into glucose. This technique does, however, have a number of drawbacks:

- the formation of undesirable by-products,

- poor flexibility (the end-product can be changed only by changing the degree of hydrolysis),
- the necessity of equipment capable of withstanding the acid used at temperatures of 140-150°C.

In all these respects, enzymes are superior to acids.

The DE (dextrose equivalent) value is used as an indication of the degree of hydrolysis of the syrup. The DE value of starch is zero and that of dextrose is 100. Syrups with DE values of 35-43 are still widely produced by acid hydrolysis despite the drawbacks mentioned above. However, due to the formation of by-products, it is difficult to produce low- and high-DE syrups of a high quality.

In the last 30 years, as new enzymes have become available, starch hydrolysis technology has been transformed. There has been a big move away from acids and today virtually all starch hydrolysis is performed using enzymes. Furthermore, in the 1970s an enzyme technique made it possible to produce a syrup as sweet as sucrose – high-fructose maize syrup. The production of this syrup has significantly boosted the growth of the starch industry in many countries, although probably more in the US than in Europe.

4.3 Enzymatic starch conversion

Depending on the enzymes used, syrups with different compositions and physical properties can be obtained from starch. The syrups are used in a wide variety of foodstuffs: soft drinks, confectionery, meats, baked products, ice cream, sauces, baby food, canned fruit, preserves, etc.

There are three basic steps in enzymatic starch conversion – liquefaction, saccharification and isomerization. In simple terms, the further a starch processor proceeds, the sweeter the syrup that can be obtained.

Firstly, there is a liquefaction process. By using bacterial alpha-amylase on its own, a 'maltodextrin' is obtained which contains mainly different oligosaccharides and dextrins. Maltodextrins are only slightly sweet and they usually undergo further conversion.

This happens during the process called saccharification. The starch already treated with bacterial alpha-amylases is made sweeter using an amyloglucosidase, otherwise known as a glucoamylase. The amyloglucosidase can theoretically hydrolyse starch completely to glucose. In practice, a little maltose and isomaltose are produced too. A pullulanase is a debranching enzyme that can also be used to aid saccharification. Fungal alpha-amylases can also be added in order to produce syrups with a higher maltose content, which means high fermentability and a relatively high degree of sweetness.

Going one step further, a proportion of the glucose can be isomerized into fructose, which is about twice as sweet as glucose. An immobilized glucose isomerase is used; without this enzyme it would not be possible to convert glucose into fructose with high yields and few by-products. In the 1970s, Novo developed the first immobilized enzyme to be produced on an industrial scale. Immobilizing the isomerase makes it possible to use it continuously for several months.

Products of isomerization that have so far assumed the greatest importance contain approximately 42% fructose/54% glucose or 55% fructose/41% glucose. These are known as 'high-fructose maize syrup', 'isosyrup', 'isoglucose' or 'starch sugar' depending on the end-use. They are as sweet as ordinary cane or beet sugar and have the same energy content. In many cases, total replacement of sugar is possible without any noticeable change in the character of the product. In the USA, for example, high-fructose maize syrup has more or less replaced the sugar previously used in the manufacture of beverages, dairy products, baked products and canned foods.

Syrups with a higher fructose content than 42% are obtained by non-enzymatic treatment of the high-fructose maize syrup. Pure fructose is about 40% sweeter than sugar.

The discovery of enzymes to convert starch into glucose has almost completely replaced chemical conversion. This most likely has led to some upstream reduction in employment in the chemical sector in favour of increased employment in the enzyme producing industry. Furthermore the discovery of enzymatic starch conversion has accelerated the replacement of sugar cane and sugar beet. Especially the discovery of an enzyme technique to produce a syrup as sweet as sucrose – high-fructose maize syrup – provided a considerable boost to the starch industry. Especially in the US it diffused rapidly into the food and drinks industry. In the EU, however, the beet growing and processing lobby was able to use EU agricultural policy to prevent high-fructose maize syrup from becoming the success it is in the US (Smith, 1993). A loss in employment in the EU agricultural sector from maize imports was prevented, particularly in sugar beet production. In the US, the success of high-fructose maize syrup drove out sugar cane imports from developing countries, leading to an increase in American employment from import substitution. Other than the import substitution effect, the replacement of chemicals by enzymes to convert starch into syrups had no perceptible effect on employment in the starch industry, due to the high degree of automation and computerisation of the production process.

4.4 Modified starches

Starch can either be sold to the food and pulp and paper industry in its native form or it can be slightly modified. Through modification it is intended to improve the properties of starch as a binder either in the food or the pulp and paper industry. In the food industry, starch is used to bind soups and sauces. In the pulp and paper industry, starch is either used in the wet process to "glue" the wood fibres together or in coatings where it provides improved gloss, smoothness and printing properties.

Raw starch is unsuitable for these applications, due to unsatisfactory flow properties. In one case, chemically modified starch with a much lower solution viscosity is used. As an economical alternative to modifying the starch with aggressive oxidizing agents, the starch can be treated with enzymes (alpha-amylases) to obtain the same viscosity reduction.

Enzymatic modification of starches is a cleaner process than chemical (oxidative) modification, as less energy is used and less waste is produced. The amount of starch ending up in wastewater is less for both types of modification as either chemically or enzymatically modified starches will attach better to the wood fibres.

The fact that enzymatic starch modification saves on energy and chemicals could possibly lead to some negative upstream employment effects in the industry producing chemicals for starch conversion and the energy sector. For the starch industry itself the switch from chemicals to enzymes is neutral in terms of employment as it only involves "a change in recipe" for the production process.

4.5 Employment effects of enzymatic starch production

Table 2 summarizes the employment effects from two enzymatic processes for altering starch: conversion into sweeteners and modified starch for paper production.

For starch conversion into glucose and high-fructose maize sweeteners, enzymes are clearly superior to the use of chemicals. Using enzymes instead of acids enables the manufacture of more precise products, including a whole range of different types of glucose. Furthermore, enzymes make it possible to achieve equivalent efficiencies in the starch conversion process starting from wheat instead of maize. The employment effects from the enzymatic conversion of starch into syrups are much more important in the US. In Europe, much depends on the penetration of high-fructose maize. It is important not to underestimate the role of EU agricultural policy in this context.

In the pulp and paper industry it is still common practice to use chemicals to modify starches. Although enzymatic modification is cheaper, it can lead to operational problems in production. Potato starch is more amenable to enzymatic modification. The choice to use either enzymes or acids to breakdown starch is therefore dependent on the sensitivity of the production process and the kind of starch that is being used. At this time, there are few employment effects from using enzymes to modify starches, as the dominant technique is still based on chemicals.

Table 2: Employment effects from enzymatic processes for starch

	Starch conversion into sweeteners (high-fructose maize syrup)	**Starch Modification**
Natural resource and extracting industries	+++ in US: substitution of maize for imported sugar 0 in EU: continued use of wheat	0
Intermediate industries		– reduced energy use
Starch industry	+ enzyme production – chemical production	+ enzyme production – chemical production – reduced energy use
Starch-using industries (food, paper and pulp)	0	0
Waste processing	– less chemical waste	– less chemical waste
Compensatory effects	0	0
Overall effect	+++ in US 0 in EU	0
Main driver	Costs	Cheaper but operational problems inhibit use

0 no effect on employment, – small reduction in employment (< 2%) + small increase in the level of employment (< 2%) ++ increase in employment between 2-5% +++ large increase in employment (> 5%).

5 Fine chemicals

The chemical sector includes the manufacture of commodity chemicals, pharmaceuticals, enzymes, refined petroleum and coal products, specialty and fine chemicals, and plastics. The manufacturing of chemicals is a major generator of materials, consumer of energy and non-renewable sources, and contributor to solid, liquid and gaseous wastes.

Biotechnology offers new ways of making chemicals that could be cleaner than current methods. Whereas bulk production of basic chemicals currently uses non-biological technologies that are so efficient that it is highly unlikely that biotechnologies could ever replace them, biotechnology is prominent in the production of fine chemicals.

Fine chemicals is one of the industrial segments where the impact of biotechnology is felt most strongly, owing to several advances in biotechnology.

First and most important, enzymes have considerable potential as biological catalysts in processes, although they are restricted to low-temperature fermentation. Whereas reactions using acids need very high temperatures, biocatalytic reactions usually take place at temperatures between 20° to 50° Celsius. As a result, however, biocatalytic processes are potentially energy-saving. Biocatalysts are also more specific and selective than their non-biological counterparts. As such they are capable of making fewer by products (specificity) and can start with less purified feedstocks (selectivity). Furthermore biocatalysts are self-propagating.

Another important feature of biotechnology in fine chemicals is its ability to produce chiral chemicals. Chirality is when a molecule has two forms that are mirror images of each other. Chemical processes usually produce both types of chiral molecules (racemic mixtures). Biocatalysis, in contrast, can produce one form or the other, or can resolve racemic mixtures, so that complicated separation processes are avoided. The preparation of pure chiral molecules is particularly crucial for the development of new drugs and pesticides, since often only one form is active.

Reactions using enzymes can often take place in water, whereas chemical reactions need harsher reaction media. The use of enzymes could lead to fewer emissions of volatile organic compounds and other harmful substances to the atmosphere. Enzymes also produce a different waste stream that can be broken down without incineration, thereby reducing CO_2 and other emissions.

Sometimes it is also possible to replace a number of chemical process steps by one single enzymatic step. A good example of this is in the manufacture of pharmaceuticals. Many pharmaceuticals are semi-synthetic molecules, in that part of their structure is synthesised by a living organism which is then modified by chemical processing. This latter part can, in some cases, be replaced by an all-enzymatic process, solving problems with product colour, the formation of by products, and low energy efficiencies.

Finally, in contrast to other industries which have traditionally relied on physical and chemical technology, biotechnology is more accepted in chemical manufacture.

Biocatalysis has substantially improved the environmental efficiency of the chemical industry. Biocatalysis represents 60% of cleaner production in this sector, while re-use and reduction of solvent use and the (biological) treatment of wastewater has also contributed to more environmentally friendly production processes. In the 1980s, biocatalysis was introduced into the production of fine chemicals and has resulted in a large reduction in waste production. Despite a four-fold increase in production volume, the production of waste was reduced by 20% through the use of biocatalysis (OECD, 1998).

Whereas the penetration of biotechnology in other user industries is quite low at this time, biotechnology has become quite important in fine chemicals. Consequently, this sector has also moved on to internalise part of the R&D. This means that the fine chemicals sub-sector is most likely to experience positive employment effects within the industry itself instead of somewhere upstream. Probably it will involve some substitution between people previously working on chemical process development and people working on biochemical processes. In the production process itself, however, there will be no significant employment effects. Similar to the other user industries, the high degree of automation and computerisation made the production process already very capital-intensive.

Table 3: Employment effects from bio-catalysis in fine chemicals

Fine chemical industry	+ R&D
	– chemical production
User industries	0
Waste processing	– less chemical waste
Compensatory effects	0
Overall effect	0
Main driver	Costs, improved quality

0 no effect on employment, – small reduction in employment (< 2%), + small increase in the level of employment (< 2%), ++ increase in employment between 2-5%, +++ large increase in employment (> 5%).

The main findings for the three industrial applications of enzyme biotechnology are given in Table 4, which lists the advantages of biological vs. chemical processes.

6 Agricultural biotechnology

6.1 Introduction

The environmental benefits of biotechnology in agriculture are due to improved crop seed varieties. These improved varieties can be produced using three different biotechnologies. The first is the use of classical breeding methods to develop new plant varieties while the most advanced type is the use of genetic engineering to achieve similar aims. In between these two methods lies assisted conventional breeding. This method combines classical breeding with several advanced technologies developed for genetic engineering, such as gene sequencing and DNA markers. Assisted conventional breeding reduces the time required to develop new varieties from approximately ten to seven years.

This case study of the environmental and employment effects of agricultural biotechnology is based on a literature review plus statistical analysis of two data sources. The first is a recent MERIT survey of European agro-seed and plant protection product (PPP) firms[3]. The second data source is the European Joint Research Council database for field releases of genetically-engineered plant varieties. In addition, interview results are available for several European seed firms.

6.2 The agro-food chain

Agricultural biotechnology can influence employment at several points along the agro-food chain, which runs from the suppliers of agricultural inputs to the final consumers of agricultural products, as shown in Figure 1, with the possible exception of retailers. Retailers can play a major role, however, by influencing agricultural prices and the types of crops that are grown.

Background rates of employment growth in the agro-food chain are determined by several factors outside the control of agbio firms. One factor is slow population growth within the European Union, which limits demand. Another important factor is the effect on farm-gate prices for agricultural crops of government policies. The March 1999 Berlin agreement to reform the Common Agricultural Policy (CAP) will alter the current price structure, leading farmers to shift from one crop type to another. The OECD (2000) predicts that CAP reform will increase European wheat production and decrease the output of oilseed, maize, and coarse grains. Although individual seed and PPP firms could gain or lose employment, it is unlikely that shifts in the types of crops that are grown will have a detectable effect on aggregate employment in the agro-food chain.

---

3 The survey and interviews were funded by the TSER project PITA on sustainable agriculture.

Table 4: Biological versus chemical processes by industrial sector

	Process	Product	Employment	Environment
Pulp and Paper				
Biopulping	- energy savings	- improved paper strength	- substitution employment chemical production by enzyme production - reduction employment energy production	- less chemicals - less CO_2 emissions from energy production
Enzymatic pitch control	- less operational problems due to pitch agglomerates - no storage of wood needed to reduce pitch	- improved paper quality	- substitution employment chemical production by enzyme production - less employment in storage of wood	- less chemicals
Enzymatic deinking	- improved deinking performance	- improved pulp cleanliness - brighter pulp	- substitution employment chemical production by enzyme production	- less chemicals
Bleach boosting	- pulp more susceptible to bleaching chemicals - possible to wash out more lignin from the pulp	- almost no loss in pulp yield and quality	- substitution employment chemical production by enzyme production	- significantly less chemicals (chlorine)
Other enzymatic applications	- improvement drainage rates - reduction processing time - reduction energy use	- enhancing pulp fibrillation - improving paper strength and quality	- increase in employment enzyme production sometimes combined with decrease employment chemical production	- less CO_2 emissions from energy production - lower BOD because of less starches in wastewater due to improved attachment to fibres

Table 4: continued

Starch industry				
Starch conversion	- higher specificity/ less by-products - lower temperatures/less energy use	- larger product range	- substitution employment chemical production by enzyme production - crop substitution in agricultural sector	- less by-products - less CO_2 emissions from energy production
Starch modification	- cheaper process due to lower costs enzymes compared to chemicals	- less homogenous product (causing problems in downstream industry)	- substitution employment chemical production by enzyme production	- less chemicals
Fine Chemicals				
- biocatalysis	- low temp fermentation – higher specificity and selectivity/ less need for pure feedstock - self-propagating - ability to produce enantiomerically pure chemicals (pharma) - water as reaction medium - less process steps possible		- increased employment within sector in R&D or substitution employment chemical process development by biochemical process development	- less CO_2 emissions from energy production - fewer by-products - less emissions of VOC - less harmful waste

Conversely, a decline in aggregate price subsidies due to CAP reform or world trade agreements is likely to reduce agricultural output and employment among input suppliers, on the farm, and in transport and distribution. Current levels of agricultural employment are not sustainable without subsidies, as shown by a 10.7% decline in the gross value-added of European agricultural output (at market prices) between 1990 and 1997. One cause of the decline is an increase in the cost of inputs such as fertilisers, pesticides, maintenance, and animal feed. The increase in input costs has largely been met by increasing subsidies from 15.4 billion to 36.7 billion €.

Figure 1: Agro-food chain

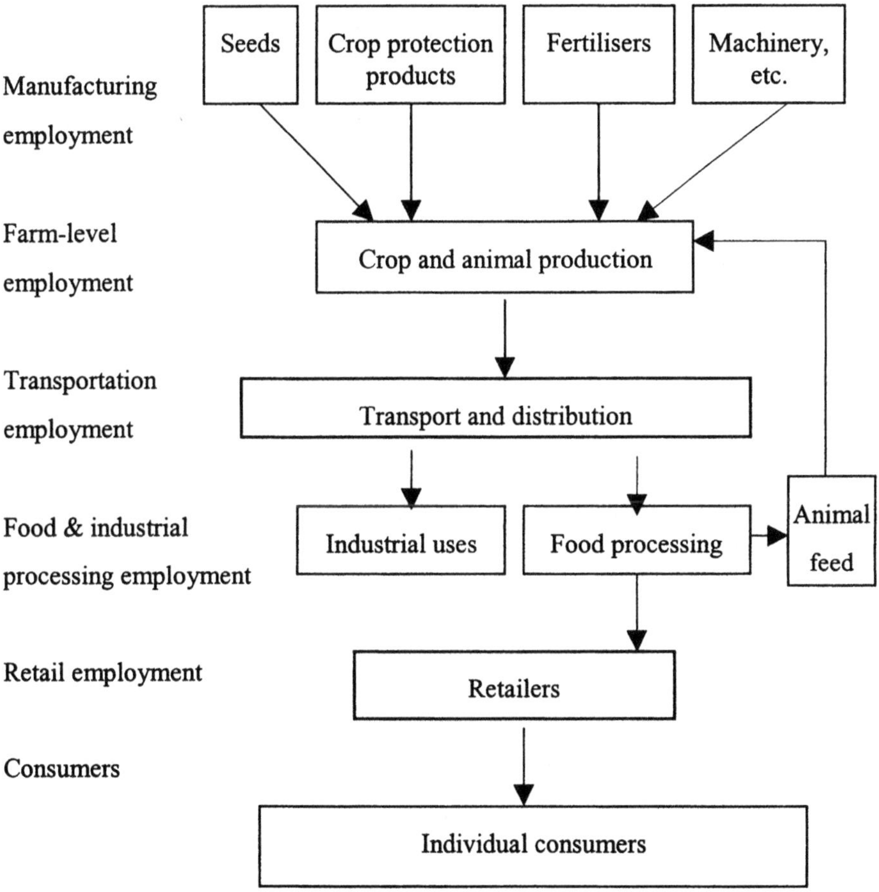

Green agricultural policies to pay farmers for land stewardship could reduce the negative effect of a reduction in price subsidies on farm level employment, but they are less likely to support employment in the seed and PPP sectors. One exception is fiscal measures to encourage farmers to adopt environmentally beneficial inputs, such as phytase-reduced feed corn[4].

6.3 Environmental benefits of agricultural biotechnology

There are two main routes through which agricultural biotechnology can lead to environmental benefits. First, new seed varieties can incorporate agronomic traits that reduce inputs of pesticides, water, and fertilisers per unit of output, or which improve tolerance to drought, cold, and salinity. Second, quality improvements can increase the production of desirable characteristics. One example is improved sugar beet that improves the efficiency of food processing, while another example is low phytase feed crops that reduces phosphate pollution from animal manure. A third example is the development of crops that can be used as industrial feedstocks or for biomass in energy production. This can result in environmental benefits if the life cycle of crop feedstocks is less environmentally damaging than that of chemical or petrochemical feedstocks (Gaskell et al., 2000).

The environmental benefits of agricultural biotechnology are considerably more controversial than the use of environmental biotechnology in industrial applications (Gaskell et al., 2000).

The debate focuses on the impacts of GM crop varieties, but some of the issues apply to all crop development programmes. This is because many of the traits, such as herbicide tolerance that have been developed via genetic engineering can also be developed through classical or assisted conventional breeding[5]. There are exceptions in which the environmental effects are limited to GM crops. These concern trans-gene GM crops in which genetic material crosses the species barrier, such as in the case of Bt-maize. Two environmental concerns are that the Bt toxin

---

4 Improved quality traits for animal feed could have substantial environmental benefits because of the size of animal husbandry in agricultural output. Developments such as low phytate maize could have strong environmental benefits by reducing water pollution, but the cost savings for the animal farmer will be very low. The problem is that low phytate maize will need to cost more than other maize varieties because of development costs and the need for identity preservation. The animal farmer is unlikely to pay the extra costs without an economic incentive, either via tax incentives, or pollution taxes on other maize varieties. Other quality traits, such as feed that is better matched to the nutritional requirements of specific species, could reduce input costs for animal farmers. The employment effect would be neutral for the animal farmer but negative for farmers that grow animal feed, due to a decline in demand.

5 For example, Monsanto used genetic engineering to develop herbicide resistant maize and soybean varieties, while Cyanamid developed herbicide resistant varieties without using genetic engineering.

could kill non-target insect species or that constant exposure to Bt toxin could result in insect pests that are resistant to Bt[6].

The environmental benefits of GM crops within a specific growing region depend on input use, for instance the amount of pesticides that are used per hectare. However, environmental benefits from a national or even a global perspective depend on inputs per unit of output. An increase in a specific input such as herbicides on a local scale could be balanced by substantially larger outputs per unit of inputs.

There is little evidence to date on the environmental benefits of currently available GM crops, which are largely limited to herbicide resistance in cotton, maize, soybeans and rapeseed and insect resistance through the insertion of the Bt gene in cotton and maize. The most relevant study is by the Economic Research Service (ERS) of the U.S. Department of Agriculture. This study used an ecological research design to analyse GM crops grown in the US in 1997.

The results, shown in Table 5, indicate that there was a decrease in pesticide use per unit of output from GM crops in all but one analysis[7], although the advantages to the farmer are less consistent.

For several GM crops, the advantages to the farmer in terms of increased yields are small and in a few cases do not cover the higher cost for GM seed. Other research has found lower yields for herbicide tolerant crops (Fulton and Keyowski, 1999, Benbrook, 1999). A review of why farmers adopt herbicide tolerant crops shows that the major reason is convenience, rather than savings from less herbicide use or higher yields (DG Agriculture, 2000b).

---

6 Both are reasonable concerns. Recent studies have shown that Bt toxin from GM crops remain in the soil for up to 200 days, which could pose a hazard to many non-target insect species. The rapid development of insect pest resistance to chemical insecticides also strongly suggests that the efficacy of pest-resistant GMs will be short lived. This would simply replace the chemical model of a continual search for new insecticides with a biotechnology model in which there is a continual search for new genes.

7 Although the evidence given in Table 1 indicates that herbicide tolerant varieties reduce total herbicide use, this interpretation depends on the comparison group, which largely consist of farmers that use conventional crop growing methods that are heavily dependent on pesticide use. The results could be rather different if the comparison group consisted of farmers that used integrated pest management techniques. This raises one of the main environmental objections against the use of GM crops with pesticide or herbicide resistance. A shift by farmers from non genetically-engineered crops to genetically-engineered crops could lock agriculture into another "one crop one pesticide model", since new genes for pest resistance will need to be continually sought to overcome pest resistance. A dependency on 'genes' would simply replace a dependency on the continual discovery of new pesticides. This could prevent greater environmental gains from other farming techniques such as integrated pest management.

6.4 Employment effects of agricultural biotechnology

There are five possible options for innovation in plant breeding (DG Agriculture, 2000a). Three of these options have potential environmental benefits[8]. For the first option of higher yields, the environmental benefits derive from a reduction in land under cultivation. The second option, higher yields per unit of output, could reduce pesticide, fertiliser and water use, even if there is no change in yields per hectare. The third option has the most diverse set of possible environmental benefits, many of which could occur by reducing inputs into industry and food processing or by reducing waste production in industry, food processing, and animal production.

Table 5: Results of ERS comparisons between genetically-engineered (GE) and non-GE cotton, soybean and maize crops in the US in 1997

Crop variety	Yield	Pesticide use	Pesticide use per unit yield
Results of econometric analyses[1]			
Herbicide tolerant cotton	Increase	No difference	Decrease
Herbicide tolerant soybeans	Very small increase	Decrease	Decrease
Bt Cotton	Increase	Decrease	Decrease
Comparison of means[2]			
Herbicide tolerant cotton	No difference	Decrease	Decrease
Herbicide tolerant soybeans	No difference	Decrease	Decrease
Herbicide tolerant maize	No difference	No difference	No difference
Bt Cotton	Increase	Decrease	Decrease
Bt Maize	Small increase	Decrease	Decrease

[1] Regression includes controls for pest infestation levels, other pest management practices, crop rotation, tillage, geographic location, differences in characteristics of adopter and non-adopter farmers. [2] Comparison between mean yields and pesticide use within specific growing regions.

To this date, farmers form the main market for GM crops, which means that the crop must provide benefits to farmers, in addition to any other benefits that could accrue to the environment or elsewhere in the agro-food chain. Farmers are un-

8 The two options with no environmental benefits are seed varieties that reduce risks for farmers and input switching innovation.

likely to adopt GM crop varieties unless the extra cost is offset by an increase in the output per unit of input costs or by higher prices per unit of output. The former can occur if yields increase or if inputs requirements decline. An increase in yields will eventually translate into lower prices, leading to a possible fall in farm employment in the absence of income subsidies. A decline in inputs without any increase in yields could maintain crop prices, but result in indirect employment declines in sectors that produce agricultural inputs such as pesticides and fertilizers.

Table 6 outlines each option and its expected employment effects through the agro-food chain (with the exception of retailers, where no employment effects are expected). The estimates assume no increase in exports and no change in current subsidy levels. Domestic demand for agricultural products is assumed to be price inelastic, except for some quality characteristics.

Two of the three innovation options with environmental benefits are likely to decrease overall employment, although there could be a shift in employment from one firm to another from substitution effects. The exception is quality traits, which could increase aggregate employment by increasing the value-added of agricultural outputs. Quality traits with industrial applications, such as the use of plant oils for lubricants or biomass for energy production, will also increase agro-food employment by shifting employment from industry. In addition, quality traits that permit European production to replace imports will increase European jobs. An example is high lauric acid rapeseed which could replace imported palm and coconut oil in detergents and lubricants. Quality traits require identity preservation, or the ability to separate improved crops from other varieties throughout the agro-food chain. Identity preservation will increase employment in transport and distribution. Quality traits will also require an increase in crop prices to cover identity preservation costs, which are between 6% and 17% of the farm-gate price, depending on the crop (DG Agriculture, 2000a). Farm level employment could also increase slightly if the price paid for improved crops increases and if farmers can capture part of the price increase.

6.5 Employment changes in seed and PPP firms

More details on employment in seed and PPP firms are available from the 1999 survey by MERIT of seed and PPP firms in six EU countries: Spain, Germany, the Netherlands, France, the UK, and Denmark. Valid responses were received from 99 firms active in developing new seed varieties and from 56 firms active in developing new plant protection products. For both surveys, the response rate was 72%[9]. In total, these firms have 13,750 employees in seeds related activities and

[9] For further details on this survey see Arundel et al. (2000) and Arundel (2001).

13,869 in PPP activities. The number of employees per firm in both surveys ranged from less than five to several thousand.

Both surveys asked similar questions on the types of technology used to develop new seed varieties or pesticides, the number of development employees, the expected change in development employees in three years, and sales and exports to non-EU countries. Table 7 provides the expected change in the number of development employees between 1999 and 2002 by the type of technology used to develop new seeds or pesticides. The three technical options for seed firms, in order of technical complexity, are conventional plant breeding, conventional assisted with advanced techniques such as gene markers or DNA sequencing, and genetic engineering. Seed firms are classified by the most technically advanced technology in use to develop new seed varieties. For example, a firm that uses both assisted conventional technology and genetic engineering is classified in the latter technology. PPP firms are classified by the type of pesticides that they develop, with three options: chemical pesticides, bio-pesticides, and chemical-crop combinations, such as herbicide tolerant maize.

The results given in Table 7 are weighted by the total number of employees in the firm, so that a firm with 1,000 employees contributes ten times more to the weighted employment estimates than a firm with 100 employees. Overall, the number of developmental employees in seed firms is expected to increase by 7.4% over three years, which is over double the expected increase in PPP firms of 3.3%. The differences by type of technology in use among seed firms are not statistically significant. For PPP firms, expected employment growth for firms that only develop chemical pesticides is minimal, at 0.7%, and highest among bio-pesticide firms, at 26.6%.

It is unlikely that the estimated changes in the number of development employees accurately predicts future employment levels. This is because the minor employment changes shown in Table 7 are likely to be completely dominated by other events, such as mergers or possible changes to agricultural subsidies. Nevertheless, the estimates can be used to predict future employment flows based on the relative change in seeds vs. PPP employment. The number of development employees is growing twice as fast among seed than among PPP firms. In the PPP sector, employment is shifting out of chemical pesticides towards bio-pesticides (albeit from a small initial employment level) and towards chemical-crop combinations. What we have is a substitution effect.

Table 6: Employment effects of agricultural biotechnology

	Increases in yield per hectare	Increases in yield per unit of inputs	Improved quality traits
Input suppliers (PPPs, fertilisers, industrial feed stocks, etc)	0 higher inputs per hectare balance a decline in farmland	– decline in demand for inputs	– decline in demand for industrial and food processor suppliers + increase in demand for PPPs for higher value-added crops
Seed firms	0 higher prices for better seed	0 higher prices for better seed	+ increased demand for seed
Farm level	– decline in hectares	0 no effect if yield per hectare unaffected	+ increased demand for crops & higher prices
Transportation	0	0	++ extra cost of identity preservation
Food processors	0	0	– lower manufacturing costs
Industrial processors	0	0	– lower manufacturing costs
Overall effect	–	–	+ agro-food chain – industrial processing
Main drivers	Cost savings for farmers	Cost savings for farmers	Higher prices through more value-added

0 no effect on employment, – small reduction in employment (< 2%), + small increase in the level of employment (< 2%), ++ increase in employment between 2-5%, +++ large increase in employment (> 5%).

The low expected growth rates for development employees in the pesticides sector needs to be viewed in terms of the long-term decline in total employment in industrial chemicals in Europe, which includes pesticide firms. Slightly positive growth rates for development employees, against a decline in overall employment, suggests a gradual shift in employment in this sector towards research positions.

Another trend among seed firms that was identified in the interviews is to concentrate breeding activities on a few core crops, or on the major industrial crops such as maize, grains, and oilseeds where large market sizes increase the potential profits and provide a greater chance of recouping development costs. This trend is

Table 7: Predicted change in development employees among seed and PPP firms in six EU countries

Development technology in use[1]	**1,999 total employees**	**1,999 total development employees**[2]	**Estimated extra development employees in 2002**	**% increase in development employees**
Seed firms				
Genetic engineering	9,405	2,308	174	7.5%
Assisted Conventional	2,488	961	54	5.6%
Unassisted conventional	1,853	404	43	10.6%
Seeds survey total	*13,746*	*3,673*	*271*	*7.4%*
Entire population estimation.[3]	*19,161*	*5,120*	*378*	
PPP firms				
Only chemicals	6,566	1,699	12	0.7%
Bio-pesticides	1,299	184	49	26.6%
Chemical and chem/ crop combinations	5,108	1,004	52	5.2%
All three types	896	288	-8	-2.7%
PPP survey total	*13,869*	*3,175*	*105*	*3.3%*
Entire population estimation.[3]	*19,318*	*4,442*	*146*	

[1] Based on the most advanced developmental technology in use for seed firms. For PPP firms, based on the types of pesticides that are under development. [2] For seed firms, includes employees active in the development or field testing of agricultural seed or plant varieties, including relevant employment in research, field testing, regulatory compliance, and management. For PPP firms, includes employees active in research, trials, and related management.[3] Crude extrapolation to the entire population of seed or PPP firms, based on the assumption that the distribution of employees is identical among 39 non-respondent seed and 22 PPP firms.

part of a general globalisation of the seed sector, but success in this strategy depends on the ability to insert high value-added traits into elite varieties that are adapted to local conditions in Europe, North America, Latin America, and Asia. Given the need to develop local varieties *in situ*, the positive employment effects among European-owned seed firms is likely to be limited to high-skilled jobs in management, administration and research. For this reason, we would expect exports to have a lower impact on employment on seed firms than in PPP firms.

This expectation is supported by the survey results, as shown in Table 8, which gives the percentage of total sales due to exports outside of the EU for seed and pesticide firms by technology type. Export rates are almost twice as high among PPP firms than among seed firms.

Table 8: Sales-weighted non-EU export rates in 1999 for seed and PPP firms in six EU countries by technology type

Seed firms		**PPP firms**	
Most advanced technology in use	**% Sales from exports**	**Type of technology**	**% Sales from exports**
Genetic engineering	20.2	Only chemicals	52.8
Assisted conventional	37.1	Bio-pesticides	52.3
Unassisted conventional	11.3	Chemical + chem/crop combinations	59.0
Average for all firms	24.6	All three types	55.3

6.6 Agro-biotechnology: towards more environmental benefits?

The current environmental and economic benefits of herbicide tolerance and pest resistance are minor, compared to the potential promise of genetic engineering. As an example, the ability to introduce nitrogen fixation genes into non-legume crops would have enormous agricultural and environmental benefits. The employment effects of quality and industrial feedstock traits could also be more substantial than that of herbicide tolerance. A basic question is when these employment effects might begin to be felt, assuming that GM crops could be planted in Europe.

This question is explored by analysing the field test data collected by the Joint Research Council of the European Commission. This dataset includes information on all field trials of genetically-modified organisms (GMOs) in the 15 EU member states since 1990, under part B of Directive 90/220/EEC.

The SNIF data contains four variables: the common name of the plant, such as 'cauliflower' or 'maize', the genetically-modified trait applied to the plant, such as 'glufosinate tolerance', the name of the company running the field trial, and the notification number, which includes information on the country where the field trial is to take place and the date of application.

For this study, 1,565 field test records were abstracted from all SNIF applications between 1990 and 2000. The database contains 62 different host species used in one or more field trials and 176 specific traits that were tested in one or more plant species. To simplify the analyses, the traits were aggregated into six major trait categories with agricultural applications: herbicide tolerance, male sterility, resistance to non-weed pests[10], industrial characteristics, quality traits, and yield. Industrial applications include plants used to produce industrial feedstocks, plus the use of crops to produce pharmaceuticals. The yield traits include stress resistance. The quality traits increase desirable properties, such as a high lysine content in soybeans. Two independent specialists checked the classification of uncommon traits to ensure that they were assigned correctly.

The 1,565 field trials test a total of 2,105 individual traits since some of the field trials are of "stacked" traits in which two or more traits are included in the same plant host. The results given here are for the 2,105 trial-trait combinations.

Our major interest is in shifts over time in the focus of investment in genetic engineering, which can be tracked using the percentage of all trial-trait combinations within each specific trait class. Currently, it takes between seven and ten years for firms to develop new plant varieties. Field trials begin two to three years into the project and can run almost until the variety is ready for commercialisation. This means that there is an up to seven years lag between the first field trials and when the variety is ready to be marketed, although recently the maximum lag should be closer to five years for most crops[11]. This lag period means that the distribution of field tests in the late 1990s indicates the types of GMO crops that are likely to be ready for commercialisation over the next five years. The analyses also indicate if investment in agricultural genetic engineering is shifting towards traits that could have more apparent environmental benefits than herbicide tolerance.

To overcome differences in the number of trials in each year, a two year moving average of the percentage of all trials due to each of the five major trait classes is calculated. Over 40% of all field trials after 1991 (which is based on very few trials) are for herbicide tolerance, followed by pesticide resistance, which hovers at just above 20% of all trials in each year. Both trends are essentially flat, showing little difference over time in the percentage of trial-trait combinations that are due to tests of herbicide tolerance and pesticide resistance.

The most important trends in terms of future benefits are for quality and industrial traits. There is a weak upward trend for quality traits, from 12.2% in 1993 to a peak of 16.5% in 2000. Industrial traits increased moderately between 1992 (4.5%) and 1998 (11.6%) but declined to 6.6% of all trial-trait combinations in

---

10 Includes insect, viral and fungal resistance.

11 Field trial permits are not required for greenhouse crops. This means that the field test may not occur until the last year or two before market commercialisation. However, greenhouse crops account for less than 20% of all SNIF trial-trait combinations.

2000. Combining quality and industrial traits results in a slight increase from 20% of trials in 1993-1994 to about 26% in 1997-1998, before falling back to 23% in 1999-2000. Overall, these results provide little support for a major shift in the near future in the types of GM crops likely to come onto the market, although there has been a slight shift towards quality traits.

Are conditions any different in the United States, where over 5,000 field trials have been conducted over the same time period? A recent study by Ditner and Lemarie (1999) analysed the American field trial data from APHIS. A higher percentage of US field trials concern pest resistance than in Europe (38.3% vs. 22.4%) while a lower percentage in the US concern herbicide tolerance (29.1% vs. 42.4%). This difference is largely due to the types of plants that are under development. These are rapeseed and beet in Europe and soybean and maize in the US. Ditner and Lemarie do not provide data on the types of traits that have been field tested over time, but they do report that the proportion of different traits is stable, with no evidence for an increase in investment in quality traits. This suggests that American research in agricultural genetic engineering, as in Europe, is not shifting notably towards traits with greater environmental benefits.

6.7 Conclusions

The results of the analysis of employment effects from agricultural biotechnology largely confirm Watanabe's (1985) prediction that the direct employment effects of agro-biotechnology are likely to be marginal and limited to R&D and production in a 'small number of highly capital- and science-intensive establishments in industrial countries'. The indirect employment effects are likely to be larger, but many of the innovations should reduce aggregate employment in the agro-food chain. In contrast, higher value-added quality traits could lead to employment increases, although public opposition in Europe to genetically-modified (GM) food could limit the development of quality traits to non-GM breeding techniques.

The field test data suggests that the indirect employment effects of agro-biotechnology in Europe is likely to be minor over the short-term of two to five years. Most of the research so far in Europe focuses on developing herbicide and pesticide tolerant varieties of major crops such as sugar beet and maize. This could slightly decrease employment among supplier firms in the plant protection products (PPP) sector, due to declines in demand for insecticides and herbicides. The effect on farm level employment is likely to be minimal, particularly as long as CAP subsidies continue to distort markets for agricultural products. The analyses of the field release data show that larger effects on employment, due to a shift towards quality traits, is at least five years away.

7 Discussion

Biotechnology is thought to be able to produce considerable environmental benefits, which is an important reason for its promotion. But what are the employment effects? And what exactly do the environmental benefits consist of? This chapter looked at the environmental benefits and the employment effects of two types of biotechnology: the use of enzymes in industry and the use of GM seeds in agriculture. Both can substitute chemicals and chemical-based products in industry and food processing.

The use of enzymes is examined for three industries: pulp and paper manufacturing, starch conversion and modification, and fine chemicals production. We find that enzyme use leads to environmental benefits in the form of less chemical waste, reduced energy use and in some cases a reduction in the use of material inputs and better product quality. Employment effects from increased enzyme use are small and the overall effect is probably negligible. The employment effects consist mainly of an upstream substitution effect in the intermediate goods industry, in which employment in chemicals manufacturing is substituted by employment in enzymes production. Employment effects in the sector of enzyme use are minimal, due to highly automated production processes. An exception is enzymatic pitch control which tends to be labour-saving compared to natural seasoning methods. There may be a negative employment effect in waste processing, due to the lesser amount of chemical waste being produced, but this effect is believed to be minor. Likewise, the reduction in employment from reduced energy use is expected to be very small given the low-labour intensity of the energy producing sector.

Diffusion has been slow, mainly because enzymes do not offer major economic benefits; learning curve effects may give some enzyme applications a competitive edge thus facilitating their widespread adoption. Currently the drivers for its use are weak.

For enzymes, employment effects will probably be concentrated primarily in the R&D stage of enzyme production, which takes place in biotech firms upstream and not within the industries themselves. Due to the high degree of automation and computerisation in process industries, switching from chemical to enzymatic processes will not have any significant impact on employment, nor will it have a perceptible effect on the skill level of the labour force. All this may change when biotechnology achieves a higher degree of penetration and gains in importance in the industry, but the drivers for its diffusion are weak. The above picture may change somewhat when biotechnology application industries internalise their R&D and biotechnology expertise, leading to increased employment. But even then the positive employment effect from enzyme use is likely to be temporary.

The environmental benefits of GM crops could be substantial. They consist of reductions in the amount of inputs, such of pesticides, water, and fertilisers, required per unit of output. There are also quality improvements in the form of higher amounts of a desirable substance. An example is high fructose maize that improves the efficiency of food processing. Another example is low phytase feed maize that reduces phosphate pollution from animal manure. The environmental benefits of agricultural biotechnology are considerably more controversial than the use of environmental biotechnology in industrial applications.

The use of GM seeds instead of plant protection products such as insecticides and herbicides does not appear to have had a notable impact on employment, although it could result in substitution effects between PPP and seed firms. Notable employment effects both in the sector of use and in other sectors (through compensatory effects) could emerge if there are large increases in yield and significant cost reductions from fewer inputs. So far, the field release data shows that very little effort is going into yield. Quality improvements hold the greatest promise for net employment increases through higher value-added crops, but there has only been a small increase in research in this area.

The value chain approach appears to constitute a useful method for assessing the employment effects throughout a value chain, and not just at the point of use. A problem encountered in this study with the value chain method was the lack of data on biotech use, and unwillingness on the part of some companies to cooperate because of public opposition to GM crops (Gaskell et al., 2000). To get sound quantitative results about the direct and indirect effects of changes in a value chain would require input-output analyses combined with scenario analysis of technology diffusion, as done by Duchin (1998). Such analyses, however, are very time-consuming and should only be undertaken to study the employment effects of major innovations.

References

Arundel, A. (1999), *Diffusion of Biotechnologies in Canada: Results From The Survey of Biotechnology Use in Canadian Industries – 1996*, Report for the Science and Technology Redesign Project, Statistics Canada, Statistics Canada Monograph No. 88F0017MPB no 6, February.

Arundel, A. (2000), Measuring the Economic Impacts of Biotechnology: From R&D to Applications, in: De La Mothe, J. and J. Niosi, *The Economic and Social Dynamics of Biotechnology*, Dordrecht.

Arundel, A. (2001), Agricultural Biotechnology in the European Union: Alternative Technologies and Economic Outcomes, *Technology Analysis and Strategic Management* Vol. 13, 265-279.

Arundel, A. and A. Rose (1998), Finding the Substance Behind the Smoke: Who Is Using Biotechnology?, *Nature Biotechnology*, Vol. 16, 596-597.

Arundel, A. and A.Rose (1999), The Diffusion of Environmental Biotechnology in Canada: Adoption Strategies and Cost Offsets, *Technovation*, Vol. 19, 551-560.

Benbrook, C. (1999), Evidence of the Magnitude and Consequences of the Roundup Ready Soybean Yield Drag from University-Based Varietal Trials in 1998, AgBioTech InfoNet Technical Paper No. 1, July.

Burke, J.F. and S.M. Thomas (1997), Agriculture Is Biotechnology's Future in Europe, *Nature Biotechnology*, Vol. 15, 695-696.

CBS (Canadian Biotechnology Strategy) Taskforce (1997), *Bio-Industries: Mining and Energy Sectors, Context and Strategic Issues Forest Sector*, CBS Taskforce, Ottawa.

DG Agriculture, European Commission (2000a), *Statistical and Economic Information for Agriculture in the European Union*, Commission of the European Communities, Brussels.

DG Agriculture, European Commission (2000b), *Economic Impacts of Genetically Modified Crops in the Agri-Food Sector*, Commission of the European Communities, Brussels.

Ditner J.M. and S. Lemarie (1999), What Can We Learn About the Development of GMOs from the American and European Field Tests Databases? Paper for the ICABR conference "*The Shape of the Coming Agricultural Biotechnology Transformation: Strategic Investment and Policy Approaches from an Economics Perspective*", University of Rome, June 17-19.

Duchin, F. (1998), *Structural Economics. Measuring Change in Technology, Lifestyles and the Environment*, Island Press, Washington D.C.

Fulton M. and L. Keyowski (1999), The Producer Benefits of Herbicide-Resistant Canola, *AgBioForum*, Vol. 2, No. 2 (Spring), http://www.agbioforum.missouri.edu

Gaskell, G., N. Allum, M Bauer et al. (2000), Biotechnology and the European Public, *Nature Biotechnology*, Vol. 18, 935-938.

Laestadius, S. (2000), Biotechnology and the Potential for a Radical Shift of Technology in the Forest Industry, *Technology Analysis and Strategic Management*, Vol. 12, 193-212.

Muller, A., G. Russel and P. Lucas (1997), *European Biotech 97: A New Economy*, Stuttgart.

OECD (1998), *Biotechnology for Clean Industrial Products and Processes*, OECD, Paris.

OECD (2000), *Agricultural Outlook 2000-2005*, OECD, Paris.

Sharp, M. (1991), Technological Trajectories and Corporate Strategy in the Diffusion of Biotechnology, in: Deiaco E., E. Hornell and G. Vickeny (Eds), *Technology and Investment: Crucial Issues for the 1990s*, London.

Statistics Canada. (1997), Biotechnology Research and Development (R&D) in Canadian Industry, 1995. *Service Bulletin Science Statistics*, Vol. 21, No. 11, 1-13, Statistics Canada, Catalogue 88-001-XPB, Ottawa.

Tils, C. and P. Sorup (1997), Biotechnology as a Cleaner Production Technology in Pulp and Paper, *IPTS Report*, Vol. 16, 5-1.

Watanabe S. (1985), Employment and the Income Implications of the 'Bio-Revolution': A Speculative Note, *International Labour Review*, 281-297.

Financial services

Neville Reeve and Steven Glynn

1 Aims of the case study

This case study will examine the case of 'green' loans, discussing the possible employment effects that they may give rise to. This will be done using the IMPRESS approach discussed in more detail below. In order to give context to the case-study there will be a brief discussion of the nature of innovation in services in general, and in financial services in particular.

The paper seeks to measure and reveal some of the facets of the relationship between 'green' innovation in financial services and employment – not just in quantitative terms but also the indirect and subtle changes that occur to the skill base that affect both innovating and non-innovating firms.

The discussion of the nature of innovation in services is intended to illustrate the difficulty in general terms of representing and understanding the process of services innovation and from observation of some empirical data to suggest what types of employment effect can occur and where they can occur in the innovation cycle. The section on UK financial services presents the economic context for the paper and reveals the importance of the sector in terms of conventional measures of size and levels of activity. Included also is a brief section looking at the broad trends in innovation in financial services. This serves as the introduction to the followings section which describes, at the macro-level, the 'greening' of financial services as a whole. A brief typology is proposed which suggests types of green innovations. Finally, in a more detailed analysis of UK financial services, actual cases of environmentally innovating companies are examined. Also presented is a profile of the main policy initiatives and services offered from some of the main financial institutions including banks, insurances companies and firms offering investment services.

All of the other case studies in the IMPRESS project address the effects of green innovation in manufacturing sectors mostly from the perspective of specific technological innovations. However, this case study – with its focus on services – offers an interesting contrast with the other work in the project and highlights the inherent problems of understanding the process and impact of services innovation. We should note particularly that a services focus requires greater recognition of organisational innovations, although of course technological innovations may also be associated with the improved delivery of a service. This is not bad thing, for a powerful criticism of theories on, and studies of, innovation has been that they have tended to put too much emphasis on technological innovation at the expense of, for example, organisational innovation (e.g. CRIC, 1999: 30).

The focus of the IMPRESS approach is on the innovation itself and the impacts that this has throughout the *value chain*. That is, the study is not limited to the impacts on the innovating firm. Consideration will be given to the wider impacts of the innovation, looking at both customers and suppliers of the innovating firm. Effects throughout the value chain that can be directly related to the innovation being studied are termed *direct effects*. It is recognised that there may be wider substitution or income effects that may occur in either another part of the innovating firm or in other firms, these are termed *indirect effects*. The case that has been chosen for study presents certain problems for this approach and these will be returned to and fully explained later in this paper.

Other work has looked at eco-services centred on the environmental service sector – that is "those providing assistance to firms confronting and/or seeking to avoid environmental problems associated with their activities" (Kastrinos and Miles, 1996: 1). Such firms include those supplying information, diagnostic services, technical problem solving, and physical environmental services. However, we propose to position this paper from a broader perspective, approaching the subject from the point of view of services in general and the environmental innovation that is taking place.

2 Innovation in services

The significant role played by services in modern economies is well-known. To take the UK as an example, services make up the largest sector of the economy in terms of both contribution to GDP and employment. While manufacturing's share in total UK GDP shrank from 24.6% in 1985 to 21.8% in 1995, the contribution of services rose from 57.3% to 65.9%. In terms of employment, the service sector has seen steady growth over the last two decades full-time equivalents in services now account for 76.2% of total UK employment.

Reflecting the increasing economic importance of services, so there has been a parallel increase in academic interest in the sector. Academic studies have highlighted the different nature of innovation in services. For example, Barras (1986) talked of a 'reverse product cycle' whereby a new technology adopted to increase the efficiency of services would eventually be used to generate new services, arguing that this paralleled the normal product cycle in the industry producing the technology. Sundbo (1997) sees service innovations as being largely market driven and formulated within the context of the firm's overall strategy. Meanwhile, Gadrey, Gallouj and Weinstein (1995) propose that the 'recombination' model can help shed light on the innovation process in services. In this perspective innovation will retain all the known characteristics of the product, recombine these characteristics, favour the reutilization of components, and may add a small difference. This clearly differs from the idea of radical innovation, and they also argue that it differs from incremental innovations, where the main characteristics stay the same and some secondary characteristics are improved or replaced.

While the usual categories of innovation – product, process and organisational – also apply to services, it is argued that other categories are also important. In particular the idea of *ad-hoc* innovation, where the innovation results from the interaction between the service provider and client is of particular interest. The high degree of supplier client interaction that is characteristic of many services can make it difficult to identify where the innovation actually takes place (Gadrey, Gallouj and Weinstein, 1995, Miles, 1996).

A further difference that has been identified regarding service innovation is that the innovation process is generally unsystematic. Services are less likely to have R&D departments and indeed, are less likely to recognise activities as R&D. This may partly explain the underperformance of service firms in statistical measure of R&D, such as the number of employees engaged in R&D, or R&D expenditure per employee.

3 Financial services

3.1 The importance of financial services

It may be useful to briefly describe some of the context for financial services, including its significance in economic and unemployment terms. Evidence has been taken from the United Nations Employment Programme, the recent Foresight work in the UK and from the Community Innovation survey of the European Commission.

By any available measure the financial services sector represents a very significant part of the European economy. As a sector it is dominated by large firms, with over 80% of workers employed in firms with more than 250 staff, and accounts for around 3-4% of employment in most EU countries. Much of the employment is high skill, white collar (around 50%). For the UK economy, the financial services sector is especially significant and it is one of the largest employers in the UK. The best estimate is that the financial services sector in the UK now accounts for in excess of 7% of GDP.

The sector is a very large user of technology with an IT expenditure more than double that for UK industry as a whole. To some extent paralleling this, the sector is employing a progressively larger number of people with skills in IT and maths. While many of these are employed in conventional roles such as IT systems design and support, others are engaged in the development of financial product and services which are underpinned increasingly by sophisticated mathematical modelling techniques (Office of Science and Technology, 1997).

3.2 Innovation in financial services

Over recent years the financial services sector has undergone rapid change and innovation. Especially important have been the results of deregulation and the introduction of new technology leading to an expansion of financial markets, a proliferation of new products, an increase in the speed of interactions, and a lessening dependence on human presence.

Much attention has been given to the sector's widespread adoption of IT. This has been seen to have a major influence on innovation in software, human-machine interfaces, telematics and related areas. IT has been employed throughout financial services as efforts are made to improve cost-efficiency. This widespread adoption of IT is perhaps reflected in the way that financial services invest particularly heavily on external technologies for innovation (CRIC, 1999: 18).

However, non-technical innovations have also been important, as is the case with services in general. A study of Danish financial services showed that technology was central to only a limited proportion on the overall innovation, namely process innovation.

Table 1: Significant innovations in the Danish financial services sector in the 1980s

	Organisational	Process	Product	Market	Overall
Non technological	94%	16%	47%	70%	54%
Not technological but dependent on technology	6%	23%	42%	30%	30%
Technological	0%	62%	11%	0%	16%

Source: Sundbo, 1997, in: CRIC, 1999: 5.

4 Environmental innovation and financial services

Whilst environmental issues did not really figure in the Foresight study of financial services, they are more explicitly recognised in the Community Innovation survey, although their importance as drivers for innovation is questioned by the results. However, actions that have been taken at both the national and international levels do reflect a growing recognition that the environment is an important area of concern for financial institutions.

4.1 The role of green issues in financial services

The Community Innovation Survey has provided some interesting quantitative insights on innovation in financial services. For the purposes of the CIS the area of 'financial intermediation' comprised the whole of Section J of NACE (Rev.1), from sectors 65 to 67, that is: 65 – financial intermediation, except insurance and pension funding; 66 – insurance and pension funding except compulsory social security; and 67 – activities auxiliary to financial intermediation.

Data from the CIS II[1] while not specifically focusing on environmental innovation, do examine a number of categories under 'aims of innovation' which can be interpreted as demonstrating some degree of environmental innovation. Clearly, these include the aims of 'reducing environmental damage', 'reducing energy consumption' and 'reducing materials consumption'. They do not however, produce a comprehensive picture of environmental innovation as other categories, such as 'open up new markets' and 'fulfil standards/regulations', may well include further environmental innovations. Focusing on financial services in the UK, the data show that while the environment is seen as relevant it is generally not seen as very important – see Table 2.

Table 2: Objectives of innovation for financial services – categories related to the environment.

Aim	Relevant	Very important
Reduce environmental damage	41%	7%
Reduce energy consumption	43%	5%
Reduce consumption of materials	46%	9%
(Open up new markets)	(99%)	(80%)
(Fulfil standards/regulations)	(87%)	(39%)

Source: CRIC, 1999.

1 Again this draws on the work done by a team co-ordinated by CRIC, University of Manchester.

4.2 United Nations Environment Programme

The United Nations Environment Programme (UNEP) has, since its inception in 1972, been concerned with encouraging economic growth that is compatible with good environmental performance. One part of this is the increasing recognition of the central role that financial services can play in the promotion of sustainable development. Working with major financial institutions, UNEP set up the Financial Institutions Initiative with the principle aim of generating debate between "all those involved in economic development and managing risks and environmentalists" (UNEP, 1999a: 3), including commercial banks, investment banks, venture capitalists, insurance and reinsurance concerns etc. In this aim the Initiative has had a degree of success with, for example, the fourth international roundtable meeting attracting more than 230 participants. While initial concerns were over risks posed to the financial sector from environmental problems, the debate has now moved on to concern risks to the environment (Kelly, 1999a: 19). A secondary aim of the initiative is to "foster private sector investment in environmentally sound technologies and services" (UNEP, 1999a: 3). In 1998 a survey was distributed through UNEP to all signatories of the Financial Institutions Initiative Statement. Highlights from the responses included that:

- 69% of respondents had an internal corporate environmental policy in addition to the UNEP statement;
- 90% of respondents had a dedicated environmental manager/department concerned with policy development and implementation, external relations and expert guidance on internal operational management;
- the majority of organisations had, or were developing, documented environmental policies covering corporate credit (74%) and project finance (63%) while fewer had policies covering investment banking (53%) or insurance (38%);
- 60% of organisations had taken steps to integrate environmental risk into credit decisions, but only 20% had formally integrated environmental risk into strategic credit or investment portfolio management;
- only a minority of organisations formally monitored environmental issues which had led to loan refusals and defaults as a separate risk classification;
- the translation of environmental impacts into financial implications was seen as the most significant obstacle to advancing the integration of environmental issues into credit and investment analysis;
- 60% of organisations had developed some form of green product (UNEP, 1999b).

5 'Green' innovation in UK financial services

5.1 Approach

Previous sections have set out ideas on the nature of services innovation and described the broad position with regard to the importance of financial services, what drives innovation in the sector and the emergence of 'green' innovation. This section attempts to go a step further and examine actual cases of 'green' innovation in financial services and explore the link to employment effects. Although the approach follows broadly the definitions of employment effects and innovation as set out for the IMPRESS study as is noted below, some modification has been necessary.

The first step in defining the scope of the work was to identify the area of focus for the work. While the importance of environmental issues for the insurance sector has been widely recognised, little work has been conducted into the effect of such issues in banking. Recent workshops commissioned by the Financial Sector Environment Forum with the support of the ESRC, highlight the growing interest in examining both the importance of the environment to the banking sector, and also the effect that banking may have on the environment. The focus on banking in this study reflects this recognised need for further study. Secondly, the types of 'green' innovation being undertaken by financial institutions, especially the major banks and lending organisations, needed to be identified. Brief descriptions of those being offered by banks and mortgage lenders are listed in Table 3. This has tended to concentrate on both the types of product innovation being offered and the 'back office' innovations in areas such as energy and paper consumption. As this case is exploring 'service innovation', it was decided that the focus would be on an actual innovation in services, rather than, for example, the introduction of new technology – hence the study of 'environmental' loans.

The next step was to explore the feasibility of applying the life-cycle approach to the analysis of new product innovations. In other words to look at the impact of new types of loan on the banks, on the organisations/firms receiving the loan, and on further downstream employment effects. For the reasons set out above, the idea of effects on suppliers in this case literally the savers with an institution has not been followed-up. Compensatory employment effects could be identified through analysis of competitors although it is very unlikely that any quantitative information would be available. Thus the boundaries for the field of analysis are:

- the banks or financial institutions offering new types of environmental lending (direct employment effects)
- their competitors in the marketplace (indirect employment effects);
- organisations receiving the loan (direct employment effects);

- the suppliers and downstream users of products or services provided by the firm that is in receipt of the loan (indirect employment effects).

5.2 The IMPRESS analytical framework

We should also note some of the practical difficulty of conducting analysis in this sector especially when trying to account for employment effects. Many of the innovations have occurred relatively recently as the wave of environmental awareness has spread across the sector. In turn, the loans made by financial institutions – that are to be a central part of this study – have also only relatively recently taken place. The effects of these, including employment gains and losses, are only now beginning to be experienced. Another point to be made concerns the difficulty of conducting an analysis of this sort in an area where confidentiality is always going to be a major issue.

Problems arise when following the IMPRESS definitions of innovation types. These concern the different categories of 'preventative' (integrated and end-of-pipe) innovation, and the boundary between these and 'curative' innovations. Take for example the issue of paper management within a financial services organisation. It is now widespread practice that many of these institutions are attempting to manage their consumption of paper products either in the internal 'back-office' operations of the bank or in the case of the paper products that are distributed to customers and as used in marketing. In many cases, targets for reducing the consumption of paper have beet set, combined with the search for alternatives. These initiatives could be categorised as 'integrated' 'preventive' measures. On the other hand, what of the attempts to deal with the excess of such paper products after they have been produced? Are these 'end-of-pipe preventive' innovations or 'curative' innovations? It seems that in such an example the distinction lies with the definition of the problem. Excess paper products which a bank has to deal with could both require an 'end-of-pipe' innovation to the process activities, and a 'curative' solution on the part of waste management authorities.

The value chain approach proposed for IMPRESS is more easily applied with some types of financial services' green innovation than others. For example, the concept of suppliers is difficult with the concept of a new environmentally targeted loan. Perhaps these suppliers are the savers that support the loan-making activity. In principle it could be hypothesised that an institution which offers 'green' loans will also attract new types of saver and new levels of saving. In turn this saving activity will, through an income effect, impact on consumption patterns. Although possible, the link would be hard to explore operationally. In the case of innovations to 'back of house' operations or service delivery, however, the notions of suppliers make more sense. Reductions in the consumption of paper for example, might have a very clear impact on suppliers of specialist papers or design services, many of which are outsourced. Downstream, all of the different types of service innovation would produce tangible effects – either on those taking-up products or, typically in the case of process innovations, on those dealing with waste products such as paper.

One of the early drivers for banks becoming involved in environmental risk assessment was the fear, raised following the Environmental Protection Act 1990, that banks could be liable for environmental damage caused by their customers. These concerns led the British Bankers' Association (BBA) to establish an Environmental Issues Advisory Panel in 1992 in order to consider the risks and to lobby government. A position statement published by the BBA in 1993 argued that banks should not be treated as 'environmental policemen' and that the risk assessment of business propositions needed to be cost effective in relation to the amount advanced. However, as has happened at the international level, this initial concern over the threat environmental problems posed to the banks has now been superseded by an interest in the role that banks can play regarding environmental performance (Cooper, 1999a: 6).

5.3 Sector survey

In order to ascertain the degree of 'environmental' action being taken across the sector an initial survey was conducted. This involved an examination of general annual reports, along with environmental reports (for those banks that produced them), see Table 3.

A taxonomy for green innovations

Taking the results of this initial survey we suggest a classification which distinguishes between innovations in operations, innovations in the delivery of services and innovations in the types of service to be developed. Innovations in operations relate to the procedures and organisation of the institutions offering the service. These could include high level changes in policy, or may be more mundane innovations located in what is often referred to as the 'back-office'. Such changes could be organisational, process, or recycling innovations. Many of this type of innovations are highlighted in Table 3. In terms of high level policy, certain banks have published environmental policies, while organisational changes have included the setting up of dedicated units to deal with environmental issues. Typically, banks and most other financial institutions are now seeking to promote their environmental credentials in the form of things such as better management of energy consumption, water consumption, use of transport, and as already noted, the use of paper. Such changes may be seen as organisational, process, or, in some cases, recycling innovations. However, as was discussed earlier, the innovation classifications are less clear here. For the financial institutions themselves there are double gains – both in the form of a better image to environmentally sensitive consumers[2] and reductions in operating costs.

The next category is innovations in the delivery of services, most of which are connected to the use of IT. Most well-known is the use of automated teller machines (ATM) and reductions in the use of cheques for cash withdrawals. The biggest environmental gains are less paper consumed, less transport of cheques and other paperwork between banks, and if the cash dispensers are widely located, fewer customer journeys to the banks. Possible environmental drawbacks might be the excess production of ATM receipts and the waste problems these present, and whatever environmental penalties might exist in the full life-cycle analysis of the manufacture and operation of ATMs. Telephone banking and electronic funds transfer facilities would be other innovative uses of IT for the delivery of services. The environmental gains and losses would be similar to those for ATMs overall less use of paper and its handling. Besides the IT innovations there are other changes, certainly linked to the power and ubiquity of IT, which have changed the relationship of the customer to the financial institutions and the services acquired. The introduction of more streamlined procedures for lending, for example either loans or mortgages, has meant fewer paper-based forms to complete whilst the customer has been able to receive approval or access to funds more quickly. Perhaps these innovations can be seen more clearly as preventative, process innovations. However, many of these changes cannot be seen as isolated innovations. In general these process changes are likely to be tied with product changes. It must be recognised that while these innovations may have environmental benefits, that is rarely the reason that they are introduced.

Finally, there are the 'green' innovations in the types of services offered by the financial institutions. In effect these are product innovations. These innovations may be introduced specifically with environmental benefits in mind, and several banks and mortgage lending institutions are now operating schemes to acknowledge the importance of loans for environmental purposes and in some case even linking these to reduced payments. For the institutions there are the promotional benefits, again in the form of appearing a sound and responsible lender. Indeed in some case banks have linked these activities to ethical policies governing the types of activity that the bank should not be engaged in. Another argument put forward is that, everything else being equal, the customer that thinks about environmental issues is thinking long-term. In banking terms, long-term means reduced risk and hence can attract better loan conditions. Of course, for other product innovations environmental benefits may just be a welcome side effect.

2 The importance of the environmental 'corporate image' is something that was highlighted in the survey aspect of the IMPRESS study.

Table 3: PREST survey of UK banks' environmental policies

Institution	Main Features of Environmental Policies
Barclays Bank	Environmental Risk Management Unit set up in 1992 to assess the impact of environmental liabilities and penalties focus on property management issues: benchmarking overall environmental performance-energy, water, use of ozone depleting chemicals environmental review of existing practices published in 1999
Co-Operative Bank	1996 environmental audit-conclusion that majority of environmental impacts are associated with purchasing commitment to the concept of the paperless office – 90% reduction in print volumes of plain paper reports-reduced transit journeys between sites ecological mission statement requires not to invest in some areas – business customers encouraged to take a pro-active stance 1994 setting up of in-house ecology unit 1995 establishment of the National Centre for Business Ecology together with 4 universities
Halifax	committed to environmental policy since December 1997, currently working on first environmental report development in 1998 of Environmental Management System – survey to identify environmental impacts – corporate environmental targets
Lloyds TSB	Group Environmental Risk Department set up in 1998 as a central point for environmental information across departments focus on property management issues: energy consumption, heating, lighting, ventilation and IT use specific attention to reducing paper consumption – search for alternative sustainable materials internal purchasing policy requires consideration of environmental issues environmental risk handbook launched 1997 to help lending officers promote good environmental practice 1999 target to promote environmental management amongst SMEs

Table 3: continued

Natwest Bank	aim to reduce energy consumption by 12% between 1996 and 2000 CO_2 emissions down by 37% since 1990 Environmental Lending Initiative relaunched in 1998 – offers loan rates below standard levels and reduced arrangement fees – open to any company that wishes to improve its environmental performance
Bank of Scotland	published an Environmental Policy in 1996 – later led to formal Environmental Management System (EMS) 1998 creation of the post of Environmental Manager with Bank's Risk Review team the EMS meant branches and departments established Environmental Action Plans based on 'best practice' environmental issues and strategy addressed quarterly by an Environmental Advisory Group some attempts to integrate environmental issues and assessments within lending approaches

As was mentioned previously, innovations in service delivery, such as the use of the internet, will be tied to product innovation, e.g. internet banking, which may have unintended environmental benefits.

In addition to the examples in Table 3, the following two cases, taken from the IMPRESS survey, provide further examples of environmentally beneficial innovations in the banking sector.

The proposed classification locates innovations according to aspects of firms' activities or operations. An alternative approach would be to describe the type of innovation itself on more conventional lines such as technological and organisational. However, as has already been discussed, it is often unclear as to where the lines are drawn. Many of the innovations described above are underpinned by organisational innovations – thus reinforcing the observation that organisational innovation is a key aspect of services innovation in general – although in several cases there would be a link to a technological outcome.

Example 1

The use of the internet to deliver products through electronic channels:

This was seen as a product, service, and distribution system innovation (it is likely that there was some degree of organisational innovation as well). Securing, and trying to increase, existing market share were important factors behind the decision to undertake this innovation. It was also seen as a way of reducing costs and improving the image of the firm. The innovation required a degree of retraining, mainly amongst distribution and sales staff and other management and office staff. The environmental benefits can be seen in the reduced material and waste costs that the bank has experienced since introducing the innovation. However, despite these benefits the innovation is not marketed as environmentally beneficial.

Example 2

The use of occupation controlled lighting, coupled with an awareness scheme delivered through the company intranet site to encourage people to think about energy-saving:

This was seen as both a process and organisational innovation. It was seen as a way of not only reducing costs but also for improving the corporate image. The innovation was cheap to undertake and did not require any retraining but has resulted in a decrease in energy costs of between 5% and 25%.

Table 4: Types of green innovation in financial services

Internal Operations	Service delivery	Products
• Environmental Management System • Targets for reductions in paper consumption and search for alternatives • Targets for reductions in energy consumption, water consumption	• IT based systems offering speedier and non paper-based financial transactions such as ATM, telephone banking • New organisational approaches, taking advantage of internal IT systems, providing simpler less paper-based procedures for those using bank services • Integration of environmental risk assessment into loan activity	• New types of financial instrument focusing on either types of customer involved in environmental activity or on innovations which have a green/environmentally sustainable aspect • Offering or privileged lending conditions for certain 'green' customers

5.4 The Co-Operative Bank

The Co-Operative Bank has been a leader in the development of environmental innovations in the Banking sector. These have involved innovation of the internal organisation, innovation in the types of services/products that are offered and innovation in terms of the Bank's relationship to external sources of specialist expertise.

The Co-Operative Bank has been in the vanguard of UK financial institutions' attempts to innovate environmentally sustainable and ethical policies and services. The Bank is a wholly-owned subsidiary of the Co-Operative Wholesale Society – a pioneer in the field of movements to integrate social values into the world of business. As at December 1998 the Bank had approximately 1.75 mln customers, supporting 2.16 mln personal accounts and 93,000 corporate accounts. Based in Manchester the Bank has a total of 3,911 employees over 66% of which work in the North West region.

In 1992 the Bank began an initiative to develop its profile on the basis of its historic positioning within the market. This meant stressing the strong ethical and moral dimension that had always supposedly been a part of what the Bank stood for. It has been out of this attempt to focus on core values that the current environmental policies and services have emerged. Further still, the environmental dimension has been intimately connected to the development of a far broader set of ethical and stakeholder values against which the Bank now sets targets and measures performance.

The first step to develop these policies was a large survey of the Bank's customers, which asked their views on the type of Bank they wanted and the types of services required. Around 200,000 questionnaires were distributed and around 20,000 responses received. One of the messages to come out was that customers wanted the Bank to develop its ethical and moral character – what was seen in effect as building on its existing brand. From the survey, an ethical policy review established an ethical mission statement for the Bank's operations and, quite significantly, the types of business the Bank did not wish to have. This was no mere paper exercise. Indeed in a few cases customers were told that the Bank no longer wished to continue doing business with them.

One aspect of the new approach was to "encourage business customers to take a pro-active stance on the environmental impact of their own activity" and to "invest in companies and organisations that avoid repeated damage to the environment" (The Co-Operative Bank, 1997a: 23).

5.4.1 The ecology unit

In 1995 the Bank established an Ecology unit – the first bank in the UK to create such a specialist department. This has evolved into what is now called the Partnership Development Team with a staff of 5. The aim for the Unit was extremely ambitious in that it would have specific responsibility for "facilitating the development of the environmental goods and services sector within the UK (The Co-Operative Bank, 1997b: 3). In practical terms this would mean promoting the development of the Bank's environmental policies and profile, working in parallel with the development of ethical policy.[3] As well as working with companies and clients of the Bank, the Unit has provided policy advice and information to those within the Bank – such as Business Centres and Relationship Managers. As well as representing a major organisational innovation in its own right, the Ecology Unit has also generated three important environmental innovations in the way the Bank does business:

- the development of the Environmental Mission Statement;
- the development of new types of loan activity offering preferential terms for work with an environmental purpose;
- the creation of a university-based centre for research and consultancy in environmental sustainability.

---

3 This was handled by the Chief Executive's department.

5.4.2 Formation of an environmental policy

The earlier work on developing an ethical policy for the Bank was extended in 1997 with the publication of an Ecological Mission Statement. Combined, these policy positions commit the Bank to:

- "Assessing our ecological impact, setting clear targets, formulating an action plan and monitoring how we meet them, and publishing the results";
- "Welcoming suppliers whose activities are compatible with both our Ethical Policy and Ecological Mission Statement, and working in partnership with them to improve our collective performance";
- "Supporting ecological projects and developing partnerships with businesses and organisations whose direct and indirect output contributes to a sustainable society"; and
- "Adhering to environmental laws, directives and guidelines while continually improving upon our own contribution to a sustainable society" (The Co-Operative Bank, 1997a: 57).

As a footnote, the bank has now gone a step further with defining its responsibilities and has brought both the environmental and ethical policy statements and commitments together in what it calls its 'Partnership Approach'. Through this, the Bank has identified seven notional 'Partners' in its operations – shareholders, customers, staff and their families, suppliers, local communities, national and international society, and past and future generations of co-operators. Three key areas of assessment have been set out – delivering value to partners; social responsibility; and ecological sustainability – and 68 measurable objectives have been defined. Each of the objectives corresponds to one or more of the identified 'Partners'. In an annually produced Partnership Report, the Bank assesses its performance against each of the 68 objectives according to fully achieved, partly achieved and not achieved.

Reflecting this environmental policy, in 1996 the Bank, along with the universities of Manchester, established the National Centre for Business Ecology (NCBE)[4]. This is an academic consultancy that provides services on a commercial basis to external clients and its sponsoring organisations. The initiative to found the NCBE came from the Bank's Ecology Unit. The result has been to provide the NCBE with commercial contacts for future work while the Bank has also contracted specific consultancy services directly from the NCBE. In parallel, the NCBE has from time to time forwarded, on an ad-hoc basis, both potential commercial business contacts and intelligence to support Bank activities.

---

4 Now called the National Centre for Business Sustainability (NCBS).

5.4.3 Environmental products and services – Greenlease

From the formation of an Ecology Unit and the development of an Environmental Mission Statement, the next stage was the development of services aimed specifically at supporting environmentally sustainable businesses. To this end preferential loans and banking arrangements have been constructed especially for those businesses which are operating in a sustainable way.

Originally it was thought that businesses would be tempted to shift their banking operations to take advantage of such loans, however, this did not work out just as planned. Relatively quickly it became apparent that the costs and uncertainties associated with such a major shift in financial affairs would not be justified by the small gain in costs that was on offer. Indeed, it is an important principle in banking that there is a fair degree of inertia when it comes to individuals' and companies' decisions to change banking operations. The solution from the Bank's point of view was to offer more targeted schemes in which businesses could shift only a small part of their banking.

Greenlease

This solution resulted in the launch of the 'Greenlease' scheme as a means through which preferential loans would be offered for companies engaged in leasing equipment that satisfied certain broad categories as being environmentally friendly. Typically the items in question would be 'off balance sheet' and thus could be handled entirely separately from the bulk of businesses operations (The Co-Operative Bank, 1997c).

Greenlease has been designed as a financial product that will help businesses make environmental improvements in a tax efficient manner over the long-term. The scheme works in the following way. Either the leasing company or the potential leaser will approach the Bank with a proposal for a loan arrangement which will then be approved according to standard banking principles of risk assessment and if the type of equipment being covered is seen to meet the environmental criteria. There are no formal arrangements to make individual assessments of the environmental pedigree of an item of technology, although the Ecology Unit has its own expertise which can be used to make a general assessment. The advantages of the Greenlease scheme are that it delivers:

- a special payment of 0.25% of the bank's investment in all capital equipment which leads to a demonstrable environmental benefit;
- access to the Bank's Ecology Unit;
- an established asset finance team who specialises in the environment sector;
- a willingness to explore projects from £100,000 upwards; and
- tailored finance packages structured to meet the requirements of the customer over three to fifteen years.

The Greenlease scheme has now been operating for around 3 years and approximately 20 loans are approved each year with a total value of £5 mln.

From the CoOp Bank's perspective the offering of specialist types of loan and other services at preferential rates can both satisfy the mission to operate according to the principles of environmental sustainability, while at the same time it can be justified against conventional banking criteria. It is important to stress that this was not seen as a charitable or other form of social activity. The rationale for offering preferential rates and charges rests on the assumption that business which attempt to operate in a sustainable way are also taking account of the long term and hence have a more deliberate strategic approach to risk and other commercial uncertainties.

5.5 Triodos Bank

5.5.1 Background

Triodos Bank, which was established in the Netherlands in 1980, was set up with the aim of supporting social, charitable and environmental initiatives. This was in response to a growing concern that enterprises should be as much value-led as designed to generate profits. The intention was that Triodos should provide financial support for such enterprises. Triodos moved into the UK in 1995 after being approached by Mercury Provident who were based in Sussex and provided a similar service to that of Triodos. Mercury Provident was too small to really operate efficiently hence the approach to Triodos which took over Mercury's operations. This type of friendly take-over has been repeated in other countries such as Belgium and has been the main route by which Triodos has expanded its operations.

5.5.2 Environmental loans

In terms of services offered, a striking feature of the Triodos operation is how it has grown in an organic fashion, for instance without a clear set of environmental priorities or services designed for particular purposes, but responding to the needs that have arisen through customers. While the Bank may have specific types of loan package an important feature is the amount of flexibility that can be achieved. Triodos does not get involved with helping customers to improve environmental performance beyond the actual loan itself.

Yet, some priority funding areas do exist. For example the support for renewable energies including a fund for wind energy that was set up in the wake of the Chernobyl disaster in 1986. Linked to this also is a savings account (associated with Friends of the Earth) in which customers know that their money is used to help fund renewable energy projects. A similar scheme operates in the organics sector which is of increasing importance to Triodos, stimulated by the debate over GM foods. The 'organic savings account' is linked with the 'Soil Association'. Triodos offer a specific 'conversion' lending package to those who wish to convert to organic farming. This is seen as a difficult undertaking as there is a period during conversion when income drops due to the farmer not yet seeing the premium from organic food. Triodos has links with Farm Research Institute and the Organic Milk Supply Co-operative (who basically guarantee premium prices) both of which can aid farmers undertaking conversion.

For a potential customer approaching the bank the first step is the referral to a loan manager who will then take the application to the Credit Committee. This considers whether the Bank can support the loan in principle. In deciding this, the Bank does not apply any strict criteria – as it is seen that an area changes too quickly. Using the expertise that has been built up by the Bank, the assessment is simply whether or not the loan would result in a positive environmental step forward. The loan has to be approved by everyone on the Committee. If the Committee doesn't feel that it has the expertise to judge the loan then they may ask for an advisor to be involved (although this is rare). This decision making process reflects the integrated style of operations within the Bank – there is no loans department, environmental department etc. There is only one graduate employee from Environmental Sciences and in general the Bank tries to recruit from a mixture of banking/financial services backgrounds and from those with an interest in environmental issues.

Triodos does not support risky initiatives and is seen as quite a prudent lender. This is reflected in its loan recovery record. From the Bank's perspective, lending for environmental purposes attracts a new type of entrepreneur – one that is not merely interested in short-term profits but takes a longer term and more sustainable approach to his operations.

Projects that get funded by Triodos would probably get funds from other sources – they are financially viable. Triodos believes its customers gain from the Bank's expertise in the area and as a consequence are listened to and "treated like human beings". The relationship is enhanced by the fact that both Bank and customers have similar social and environmental priorities. The Bank also offers competitive rates and will match any written quotes from other banks. That customers feel well treated is critically important for the process of word-of-mouth promotion. In general, the Bank does not have the resources to promote its services directly.

The Bank does not keep any estimates of employment effects from its loan activity, although employment would typically be one of the things considered when assessing a loan application in the context of the societal benefits that would result if it were funded. Soft, rule of thumb estimates from the Bank suggest that there is probably an overall positive effect. For example organic farming – one of the important growth areas – has tended to be more people intensive than conventional intensive methods.[5] There is also the 'opportunity cost' effect. With the agricultural sector generally seen to be struggling, for many farmers, converting to organic operations has been the way to stay in business and hence preserve existing jobs. The effect is magnified given the higher than average levels of unemployment in many rural areas.

6 Concluding remarks

6.1 General remarks

This paper has demonstrated that the analysis of environmental innovation in services can be quite problematic. What is clear is that the value chain approach proposed for IMPRESS is more easily applied with some types of financial services' green innovation than others. One problem is the idea of looking at the impacts on suppliers and customers of a green innovation. For example, while reductions in the consumption of paper might have a very clear impact on suppliers of specialist papers or design services, many of which are outsourced, for the case of a green loan the idea of the suppliers is harder to pin down. Perhaps these suppliers are the savers that support the loan-making activity. In principle it could be hypothesised that an institution which offers 'green' loans will also attract new types of saver and new levels of saving. In turn this saving activity will, through an income effect, impact on consumption patterns. The link, although possible, would be hard to explore operationally.

Regarding impacts on customers, while certain cost savings resulting from the types of 'back office' innovations that have been described may be passed on the customers, it would seem unlikely that this would have a large effect. However, green loans may have a very large impact on their recipients. Even the concept of whether an innovation is environmentally beneficial becomes unclear here. Certain innovations, e.g. energy conservation may bring obvious direct environmental benefits to the firm in terms of less fossil fuel use. However, while we have argued that green loans are an environmental innovation, they do not bring any direct environmental benefit to the bank – they have more indirect environmental benefits that are passed on to the customers.

---

5 In discussions one case was highlighted where a loan was given to set up an organics operation in an area where it was difficult to create employment.

6.2 The impacts of 'green' loans

The decision was made to focus on banking, reflecting an interest in what is a rapidly changing sector. Rather than examining the impacts of a technological innovation in the banking sector, it was decided that the study would focus on an actual service product innovation, hence 'green' loans.

As has already been said, the concept of suppliers is difficult to envisage here. It is possible that employment impacts could be seen at a number of points in these value chains: the bank may increase employment; competitors may suffer; the firms receiving the loan may need to hire new people; this could have a knock on effect to the suppliers and customers; it may give rise for an increased need for environmental expertise etc. However, the work that has been carried out tends to suggest that they are likely to be small.

The bank itself may see some employment increases, such as with the establishment of a dedicated environmental unit e.g. the Partnership Development Team at the Co-op Bank. However, even for the Co-op case, which is probably the most developed, this team is only five strong. In many other banks environmental responsibility may fall to one individual. So far as the impact of green loan schemes and the like on competition between the banks, this is also not likely to be significant if taken just at its most direct. The overall level of business activity for such initiatives has been relatively small and there is no evidence that any of the organisations now operating such schemes have been able to build a leading position at the expense of the other organisations in the field. This may however be an undercounting of the true impact of these measures. In the cases studied at least, environmental loans and an environmental profile for the Banks are seen as important parts in the general market positioning of the institutions. The environmental dimension is seen as just facet of a complex profile and may not be separable in terms of quantifiable effects from other parts of the profile.

Perhaps the most obvious place to look for the impacts of 'green' loans is in the recipient organisations. In thinking about this, the question has to be addressed as to whether the organisation would have received a loan from another source. Banks such as the Co-op and Triodos argue that they only make loans on the basis of sound financial criteria, hence a proposal has to be financially viable and, if this is the case, could perhaps have received funding from elsewhere. However, while the preferential rates and specialised loans may be seen as financially sound, it is likely that they may encourage applicants who would not be able to meet a different set of repayment criteria. Hence, it could be argued that the loans are likely to have direct, positive effect on employment by allowing schemes to be undertaken that would not be possible otherwise. This could then have a knock on indirect effect of benefiting the suppliers of these recipient organisations and environmental consultants who are advising firms. However, any such effects are likely to be small and difficult to measure, as the loans are small-scale, there are not that many 'green' loan schemes, and those that there are have only been in operation for a short time. One final thing that is worth noting is that while the employment effects of these 'green' loans may be small, it is difficult to envisage it being negative.

References

Barclays (1999), *Environment Review*, April.

Barras, R. (1986), Towards a Theory of Innovation, *Research Policy*, Vol. 15, 161-173.

Cooper, R. (1999), Banking and the Environment, in: Coulson, A.B. (Ed.), *Views on the Environment: Workshop Report*, commissioned by the Financial Sector Environment Forum, January.

CRIC (1999), *Analysis of CIS Data on Innovation in the Service Sector: Interim Report No.2*, April.

Gadrey, J., F. Gallouj and O. Weinstein (1995), New Modes of Innovation: How Services Benefit Industry, *International Journal of Service Industry Management* Vol. 6, No. 3, 4-16.

Halifax (1998), *Environmental Review*, January.

Kastrinos, N. and I. Miles (1996), *Environmental Services and European Regulation – End of Award Report: Summary of Research Results*, December.

Kelly, M. (1999), United Nations Environment Programme Financial Institutions Initiative, in: Coulson, A.B. (Ed.), *op cit.*

Lloyds TSB (1999), *The Community and Our Business*, April.

Miles, I. (1996), *Innovation in Services: Services in Innovation*, Manchester Statistical Society, presented 20th February.

Office of Science and Technology (1997), *Technology Foresight – Report 3, Financial Services*, HMSO, London.

Sundbo, J. (1997), Management of Innovation in Services, *The Services Industries Journal*, Vol. 17, No. 3, 432-455.

The Co-Operative Bank (1997a), *The Partnership Report: Seven Partners, a Balanced View.*

The Co-Operative Bank (1997b), *The Clean and Efficient Choice for the Environment sector*, The Co-Operative Bank, EU1/4/97.

The Co-Operative Bank (1997c) *Greenlease – The Clean and Efficient Choice for the Environment sector*, The Co-Operative Bank, CBM108 05/97.

UNEP (1999a) *Financial Services and the Environment: "Questions and Answers"*, January, http://www.unep.ch.

UNEP (1999b) *Report of the UNEP Financial Institutions Initiative 1998 Survey*, http://www.unep.ch.

Part III: National Survey Results

The Swiss survey

Najib Harabi and Annette Jochem

1 Description of the sample

The basis of the investigation is a sample of 201 Swiss firms. About 76% of these firms have between 50 and 199 employees, 24% have more than 200 employees. It was further determined that 75% of the firms are located in German-speaking Switzerland, 21% are located in French-speaking and 4% in Italian-speaking Switzerland.

The distribution of the industries selected is similar to the distribution of the other countries' surveys and corresponds to the structure of the Swiss economy. The manufacturing industry is represented with a share of 43%. It is followed by wholesale (17%) and construction industry (11%). The agricultural sector, the public sector and the health sector have been excluded from the survey. Table 1 shows the detailed figures that describe the sample.

2 Environmental innovation

Figure 1 shows all the innovations having an intended or not intended positive impact on the environment which have been introduced by the firms in the last 3 years (column 'Mentioned'; more than one answer was possible). Among these, the innovation which has been cited as the most environmentally beneficial one is also shown (column 'Most important'). A broad spectrum of environmentally friendly innovations has been introduced in the last 3 years. Process innovations predominate among both the innovations mentioned (76% of firms; see left column) and the most environmentally beneficial innovations (43%; see right column). They are followed by products or services innovations, respectively, recycling innovations and pollution control technologies ("end-of-pipe").

Table 1: Description of the sample

	Number of firms	Share
Small (< 200 employees)	153	76
Large (>= 200 employees)	48	24
Industry/manufacturing	112	56
Service	89	44
Manufacturing	87	43
Electricity, gas and water	2	1
Construction	23	11
Wholesale/retail trade	35	17
Hotels and restaurants	8	4
Transport, storage and communication	15	8
Financial intermediation	18	9
Real estate, renting and business activity	13	7
German-CH	150	75
French-CH	43	21
Italian-CH	8	4

Figure 1: Environmentally beneficial innovation

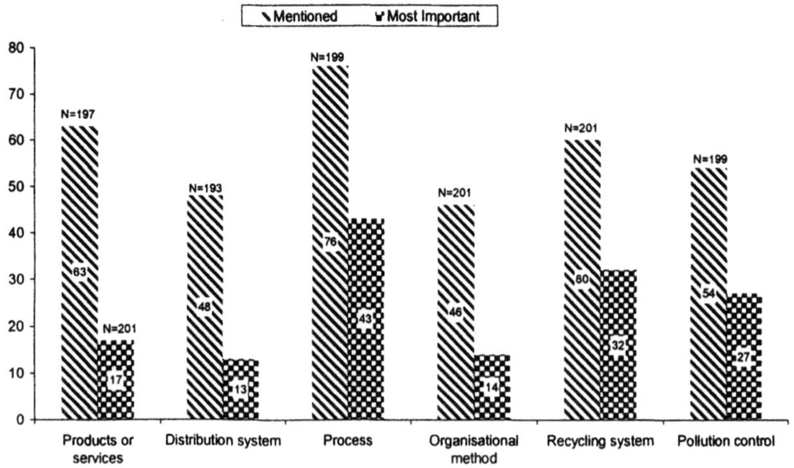

Figure 2 shows important reasons for introducing the eco-innovation, by innovation type. Improving the firm's image is one of the reasons for introducing the innovation which was mentioned most frequently, especially in those firms that had introduced service innovations and product innovations. Complying with environmental regulations was cited as the most important reason when end-of-pipe technologies were introduced. It also plays an important role for the other innovations, especially recycling innovations and product innovations. Another reason that plays a relevant role, especially for improved delivery, transport or distribution systems and for process innovations, is the reduction of costs.

All in all, the reasons for introducing the eco-innovation do not seem to depend too much on the innovation type because the columns in figure 2 differ more by reasons than by innovation types. Therefore, it can be concluded that for all the innovation types, one can build the following ranking of reasons mentioned by importance: Improving the firm's image, complying with environmental regulations, reducing costs (multiple answers were possible).

Figure 2: Important reasons for introducing the innovation

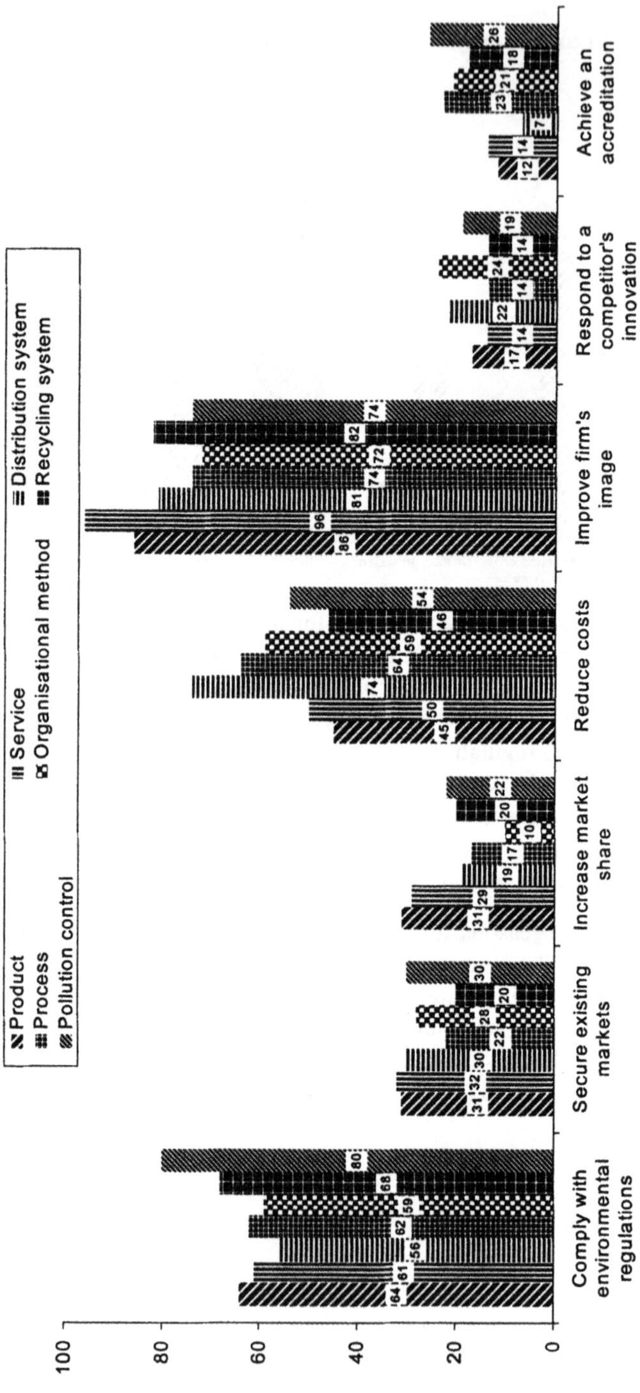

Figure 3 shows the effects of the environmental innovation on employment. 91% of the firms had no notable effect on employment due to the specific innovation (see left column). In 11 cases (5.5%) the number of long-term employees increased due to the innovation, in 7 cases (3.5%) it decreased. This clear result shows that there is no strong connection between the introduction of environmental innovations and employment.

Regarding the distribution of employment effects by innovation type, it becomes apparent that product innovations and service innovations have a sizeable above-average positive employment effect (12% and 11%). The same applies to end-of-pipe technologies (9%) while the employment effect of improved delivery, transport or distribution systems is extremely negative (11%) and by no means positive. It is further noticeable that the employment effect of recycling innovations is positive in all cases. Recycling innovations create a new link in the value added chain and thus have a potential for additional employment.

Figure 3: Effect of innovation on employment

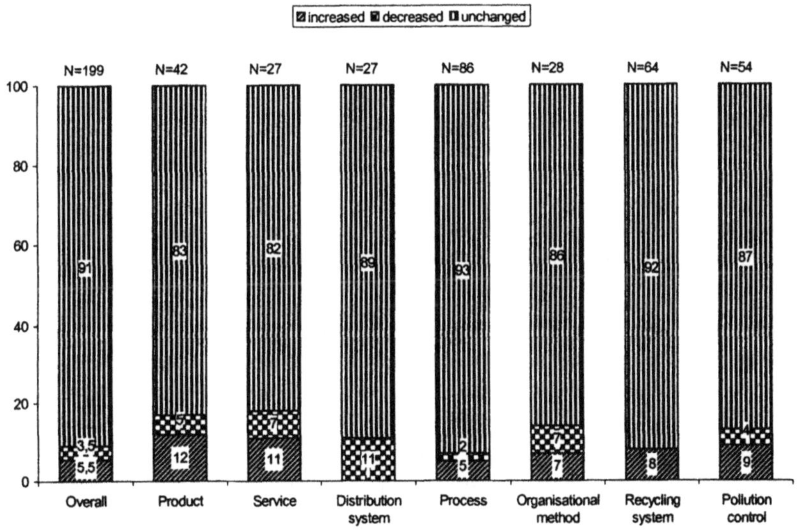

Figure 4 shows by whom the eco-innovation was developed, overall and in connection with the question whether the innovation resulted in employment effects. For 45% of the firms, other firms or institutes developed the innovation (see left column). Against this, only 24% developed the innovation in their own institution. It is surprising that those firms which decreased the number of employees usually developed the innovation themselves (57%). However, no great importance should be attached to this result because it refers to a number of only 7 firms which reduced employment. For the 180 firms which did not change their employment due to the eco-innovation, the distribution is quite similar to the overall distribution. This means that the innovation was usually developed by other firms or institutions.

It can be concluded that an additional positive employment effect can be expected from the other firms or institutions that have developed the eco-innovation.

Figure 4: Who developed the innovation

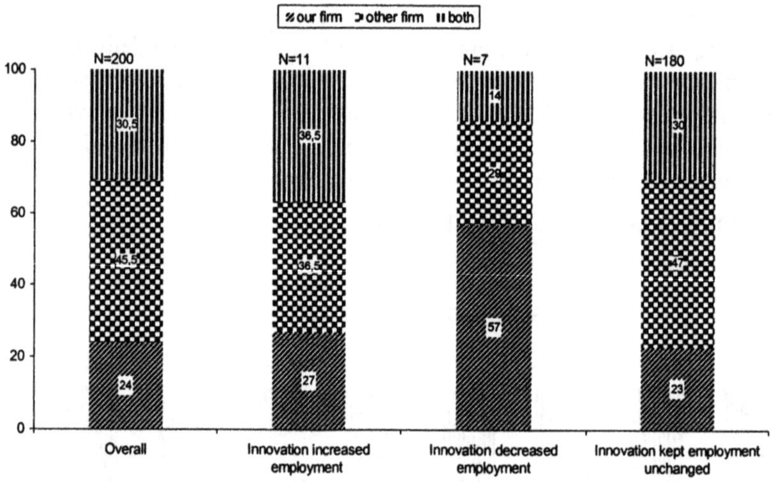

Figure 5 gives an overview of the investment costs and their correlation with the employment effect of the eco-innovation. About half of the firms invested less than € 50,000 (51%; see left column). The overall distribution of the share of investment costs is quite similar to the distribution of the investment costs for those firms which kept employment unchanged. This means that the level of investment costs is generally small. However, in the 10 firms which increased employment there is an above-average high level of investment costs. From this can be concluded that the level of investment costs has a positive influence on employment. However, firms which decreased employment due to the innovation also have an above-average high share of investment costs above € 50,000. However no great importance can be attached to this result because the number of firms having increased or decreased employment is small.

Figure 5: Investment costs of the innovation

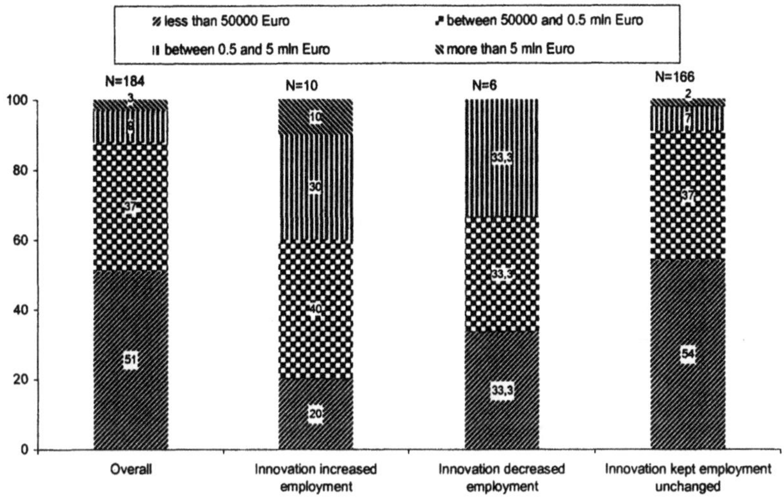

Figure 6: Proportion of innovation expenditures

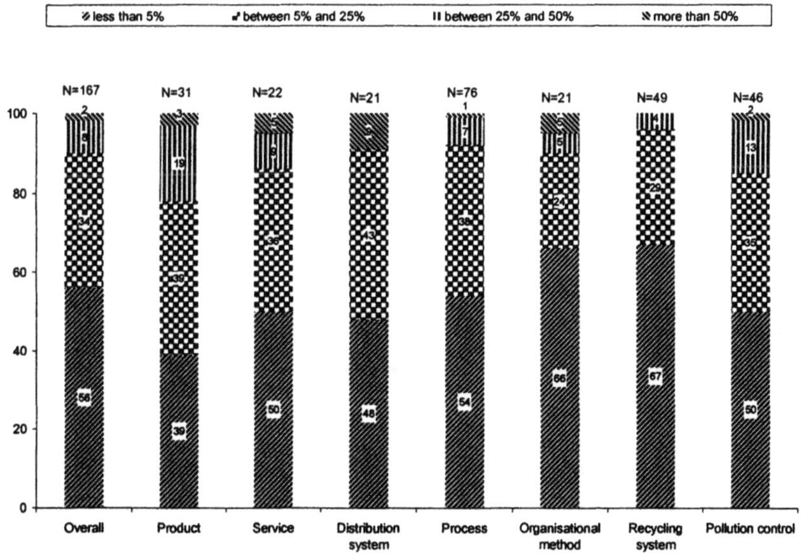

Figure 6 shows which proportions of the establishment's total innovation expenditures over the past three years were spent on the environmental innovation. For the majority of establishments (56%), less than 5% were spent on the eco-innovation. Analyzing the results by innovation types, it is remarkable that an above-average number of firms had a relatively high investment share for product innovations, distribution systems and pollution control technologies. When comparing this result with Figure 3, there is no clear connection with the employment effects of such innovation types. Product innovations and pollution control technologies have a small, above-average positive effect, but distribution systems have an extremely negative effect. This means that there is no clear connection between the proportion of innovation expenditures and the employment effect.

Figure 7 gives an answer to the question whether there is a correlation between receiving subsidies or grants for introducing the innovation and the innovation type. Generally, only 5% of firms received a subsidy or grant. For product and service innovations, an above-average number of firms received subsidies or grants. No subsidies were received for distribution and organisational innovations.

When the subsidies are correlated with the employment change due to the eco-innovation, it is surprising that neither firms which increased nor firms which decreased employment due to the innovation received subsidies. It can be concluded that subsidies given for environmental innovations have no influence on employment.

Figure 7: Was subsidy/grant received (by innovation type)

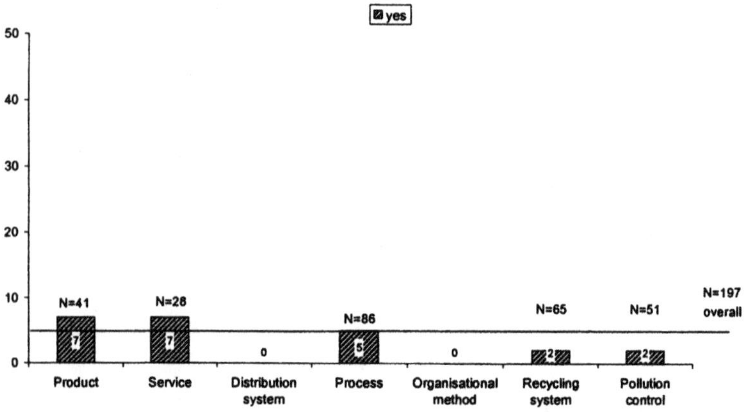

Figure 8 gives an answer to the question whether new skills required due to the eco-innovation are correlated with the innovation type and with the employment effect of the same innovation. Overall, new skills were required by about 28% of the firms introducing environmental innovations. It can be shown that service innovations (50%) as well as distribution systems (38%) require a noticeable above-average need for new skills. Furthermore, there is a strong connection between the need for new skills and a positive employment effect: 73% of firms with new long-term employees due to the innovation realized a need for new skills. On the other hand, there was also an above-average need for new skills in firms which reduced employment due to the innovation. One possible explanation for this result could be that many firms requiring additional skills outsourced the new tasks.

Figure 8: Need for new skills by innovation type and employment effect

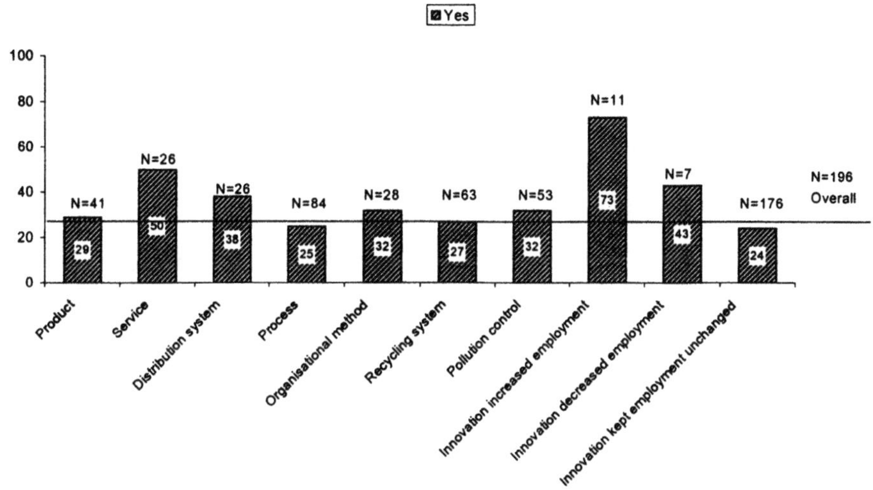

Figure 9 gives an overview of how the need for new skills was met (multiple answers were possible). The majority of the firms (90% of firms requiring new skills; 25% of all firms) uses the method of training existing employees. Only 10% of all firms (33% of firms requiring new skills) hire new employees on a permanent or temporary basis. From the 5% (17% of firms requiring new skills) which outsourced this work, it can be expected that positive employment effects occurred in the firms commissioned.

Figure 9: How were new skills met

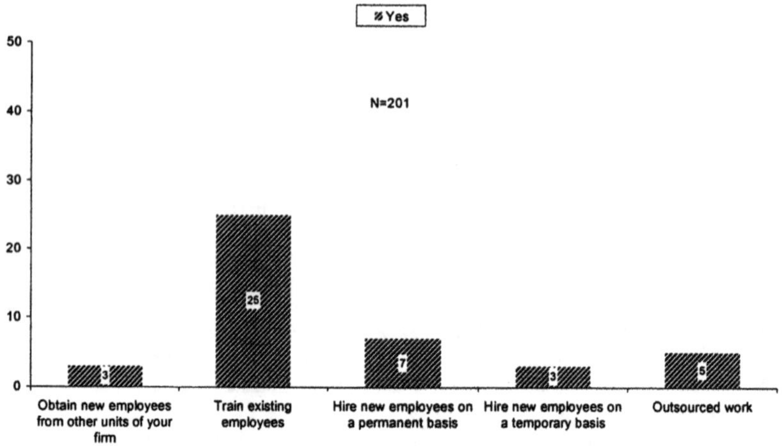

Figure 10 gives an answer to the question who received training and whether there is a connection with manufacturing or service firms or with the employment effect. Overall, mostly skilled workers received training (22%), followed by staff with college or university degrees (8%) and by unskilled workers (3%). This distribution is similar for both manufacturing and service firms. It is also similar for firms which increased employment and for firms which decreased employment. The high share of skilled workers can be explained by the innovations introduced. Mostly not having developed the innovations by themselves, the firms meet their demand for employees and skills for application but not for the development of the innovations.

Figure 10: Who received training by industry and employment effect

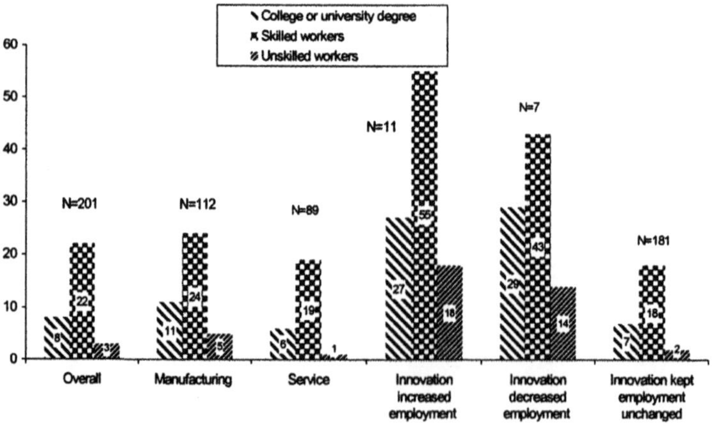

Figures 9.11a, 9.11b and 9.11c show the effects of the environmental innovations on sales, prices and costs. For more than 85% of the establishments, the innovations had no effect on sales and prices. For 14% of the establishments, sales even increased. In no cases did they decrease. Prices increased in 7% of the firms but mostly by less than 5%. Prices could be decreased in 5% of the firms (see Figure 11a).

Figure 11a: Effects of innovations on sales and prices

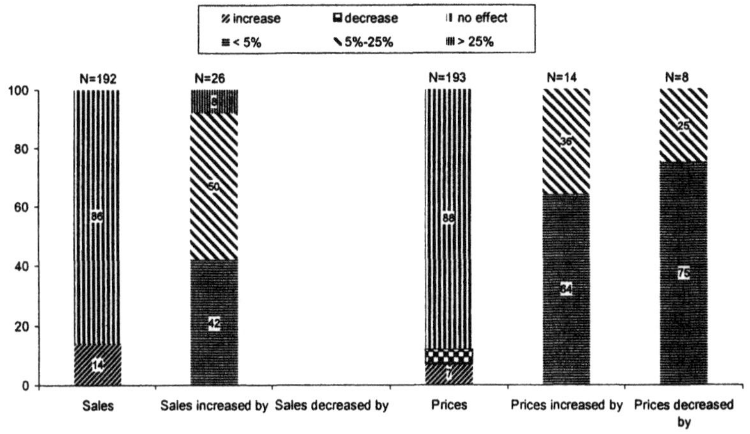

The energy costs decreased in more firms than the other costs forming part of this investigation. 43% of the firms could decrease their energy costs due to the innovation. This result goes in line with the high share of process innovations (see Figure 1) and the motive of cost reduction (see Figure 2). The level of decrease of energy costs also is relatively high: for 56% of the firms, energy costs decreased by more than 5%.

Material costs were unaffected for 63% of the firms. In those cases with changes of material costs, the number and level of decreases were a bit higher than the number and level of increases (see Figure 11b).

Figure 11b: Effects of innovations on energy and material costs

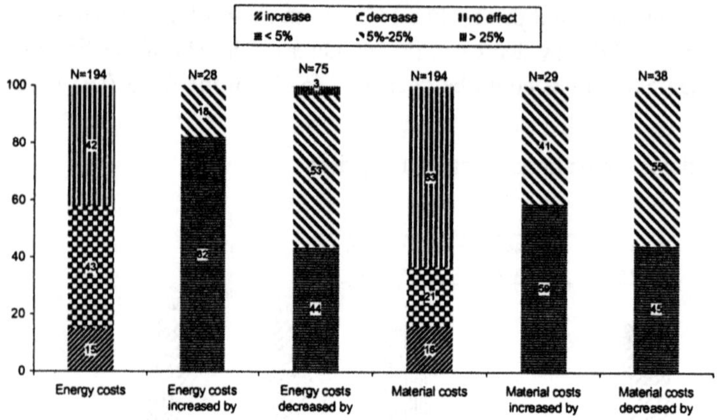

Waste disposal costs could be decreased by about one third of firms (see Figure 11c). This is due to the relatively high share of recycling innovations. The level of decrease was even more than 25% for 12% of the establishments. However, waste disposal costs increased for 20% of the firms.

In 65% of the firms, labour costs did not change due to the innovation. However, they increased for 21% of the firms but mostly by less than 5%. Compared to the increased or decreased employment (see Figure 3), the higher percentage of firms with increased or decreased labour costs can be explained by the fact that labour costs had increased due to the need for new skills. Another reason could be that in this survey, only the change in the number of long-term employees was taken into consideration. The number of short-term employees possibly changed, too.

Figure 11c: Effects of innovations on waste disposal and labour costs

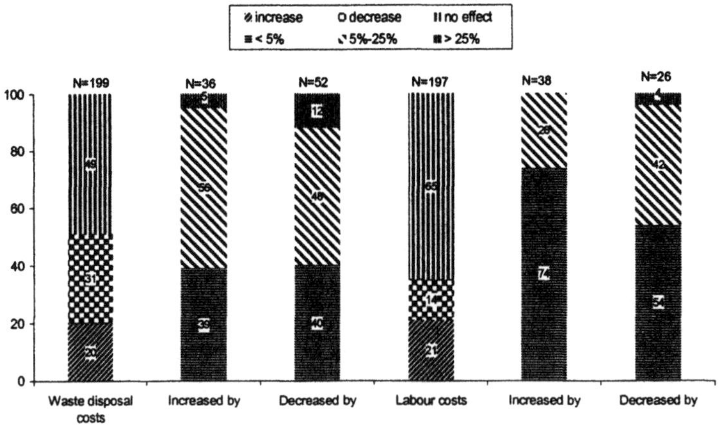

3 General questions

Figure 12 gives an answer to the question how important environmental regulations are for processes and products and whether concrete changes were made in the last three years in order to comply with these regulations. It becomes apparent that environmental regulations seem to be important for both product and process innovations. Concrete changes in order to comply with the regulations were made in about half of the eco-innovative firms

Price and quality can be clearly defined as the most important competition factors (see Figure 13). By a wide margin, they are followed by innovative products or services, corporate image and environmentally friendly features.

On the one hand, it is not surprising that quality is the most important competition factor in Switzerland which has one of the highest living standards among all industrial countries. On the other hand, environmentally friendly features are mentioned by only 2% of the firms as the most important factor which is surprising because the answering firms are exclusively those which introduced environmental innovations in the last three years..

Figure 12: Importance of environmental regulations for processes and products and changes to comply with regulations

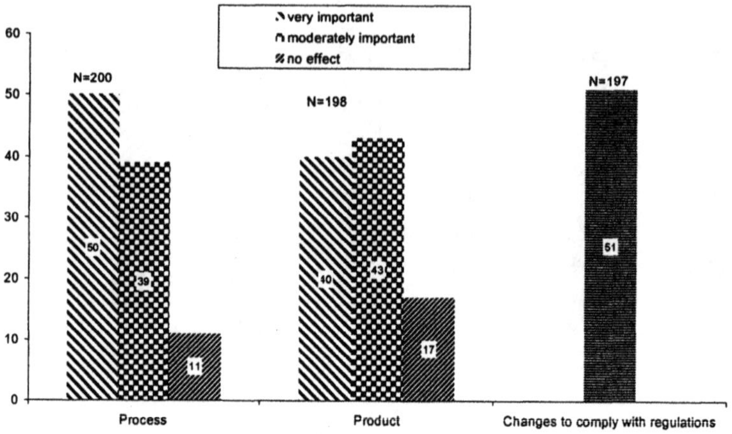

Figure 13: Important competition factors

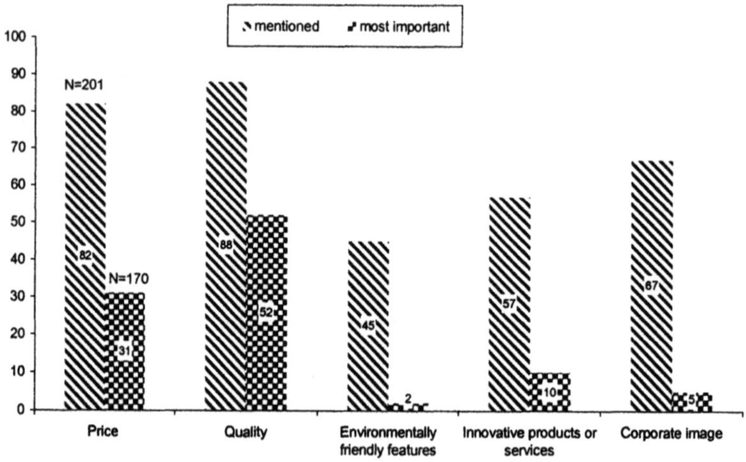

Figure 14 shows the overall employment changes of the firms introducing environmental innovations. For the majority, employment increased in the last three years (46%; see left column). This shows that also environmentally friendly firms follow the general trend of employment increase in Switzerland.

Figure 14 furthermore shows that more large than small firms, and more service than manufacturing firms increased employment. This corresponds to the fact that service firms generally gain employment share in the economy. The distribution of the overall employment change can also be applied to those firms which did not change their number of long-term employees due to the environmental innovation (see right column).

Figure 14: Change in overall employment

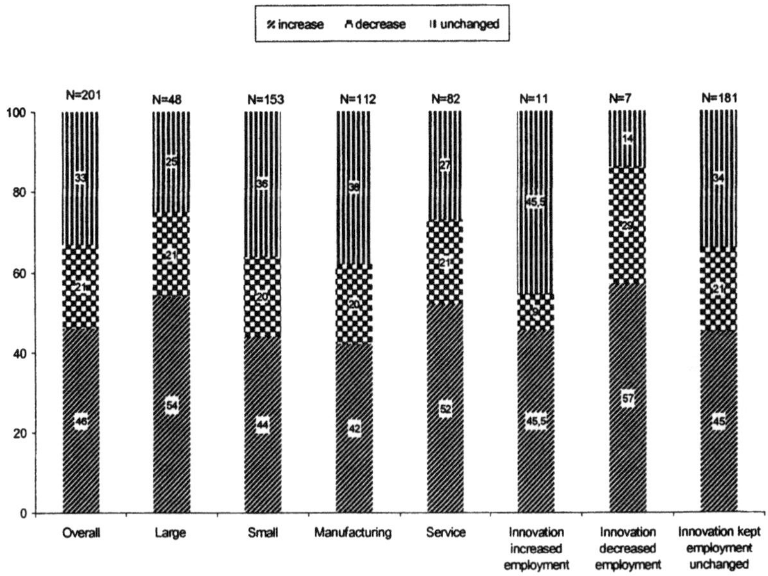

Figure 15 gives an answer to the question whether there is a correlation between employment change due to the eco-innovation and those firms which offer environmentally friendly products or services. There does not seem to be a strong connection because the percentage of those firms which kept employment unchanged due to the innovation is about the same as that of the firms offering environmentally friendly products. Due to the small number of firms having increased or decreased employment, no general conclusions can be drawn from these two groups.

When comparing the firms offering environmentally friendly products with the overall employment change in the establishments in the last three years, no strong connection can be seen between these two variables either.

Figure 15: Does your company offer products or services that are specifically marketed as environmentally friendly

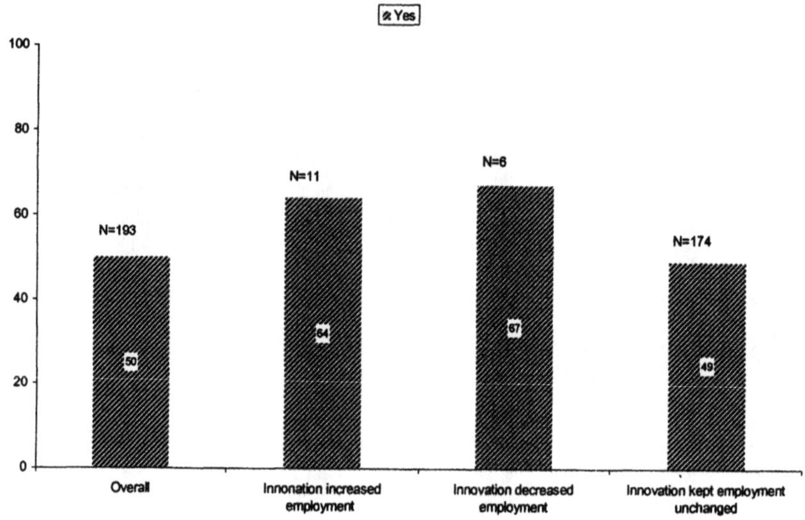

4 Conclusions

The Swiss survey comes to the result that in general environmentally beneficial innovations have no sizeable effect on employment. However, regarding the small number of firms that have changed their number of long-term employees due to an eco-innovation, there are more firms that increased rather than decreased their number of employees. In addition, it can be concluded that the positive employment effect, entailing increasing labour costs, had no negative impact on sales even if prices often had to be raised. The positive influence on the image of the firm and the quality of the products seems to compensate for the higher prices.

In addition, a further positive employment effect can be expected from the tasks outsourced due to the eco-innovation. The effect cannot be quantified in this study, however. Also other indirect employment effects such as those resulting from the firms that had developed the eco-innovations cannot be quantified.

Besides the employment aspects, many firms were able to decrease prices because other important costs like energy costs and waste disposal costs declined considerably due to the eco-innovation.

The German survey

Klaus Rennings and Thomas Zwick

1 Description of the sample

The basis of the investigation is a sample of 401 firms. About 78% of these firms have between 50 and 199 employees, 22% have more than 200 employees.

The distribution of the industries selected is similar to the distribution of the other countries. The manufacturing industry is represented with a share of 44%. It is followed by wholesale (24%) and renting and business activity (12%). The agricultural sector, the public sector and the health sector have been excluded from the survey. Table 1 shows the detailed figures that describe the sample.

The German sample has been also specified in terms of geographical location of the firms considered. The distincion between West Germany and East Germany has been set according to the relevant structural differences between the two geographical areas. 64% (unweighted) of the establishments are located in the west and 36% (unweighted) are located in the east of Germany.

Table 1: Description of the sample

	Number of firms	Share	Weighted share
Small	317	79.05	77.63
Large	84	20.95	22.37
Industry/manufacturing	219	54.61	53.24
Service	182	45.93	46.76
Manufacturing	169	42.14	43.66
Electricity, gas and water	8	2.00	1.13
Construction	42	10.47	8.28
Wholesale/retail trade	89	22.19	24.18
Hotels and restaurants	10	2.40	2.19

Table 1: continued

Transport, storage and communication	31	7.73	7.73
Financial intermediation	2	0.05	0.66
Real estate, renting and business activity	50	12.47	11.99
West Germany	258	64.34	84.55
East Germany	143	35.66	15.45

2 Environmental innovations

Figure 1 shows all the innovations having an intended or not intended positive impact on the environment which have been introduced by the firms in the last 3 years (column 'Mentioned'; more than one answer was possible). Among these, the innovation which has been cited as the most environmentally beneficial one is also shown (column 'Most important'). A broad spectrum of environmentally friendly innovations has been introduced in the last 3 years. Process integrated innovations predominate among both the innovations mentioned (72% of firms). They are followed by product or service innovations, respectively, and by recycling innovations.

Figure 2 shows important reasons for introducing the eco-innovation, distributed by innovation type (multiple answers were possible). The three most cited reasons for introducing the innovation are to improve the firm's image, to comply with environmental regulation and to reduce costs. Market goals play only a minor role.

Figure 3 shows the effects of the environmental innovation on employment. 83% of the firms had no notable effect on employment from the specific innovation (see left column).

Figure 1: Environmentally beneficial innovation

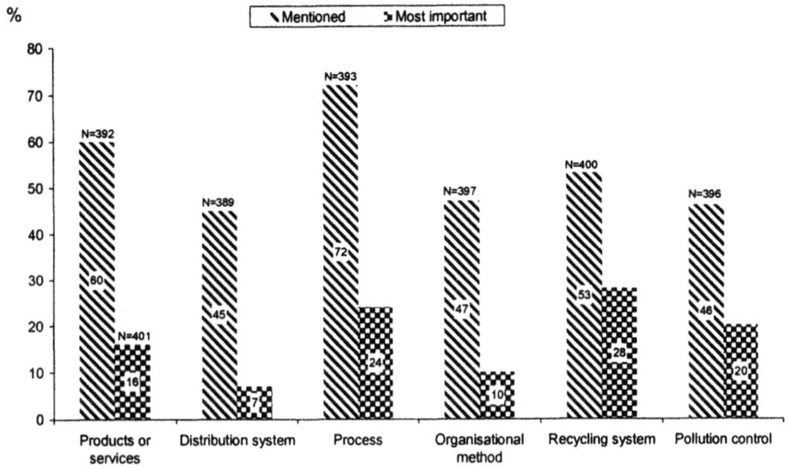

Figure 2: Important reasons for introducing the innovation

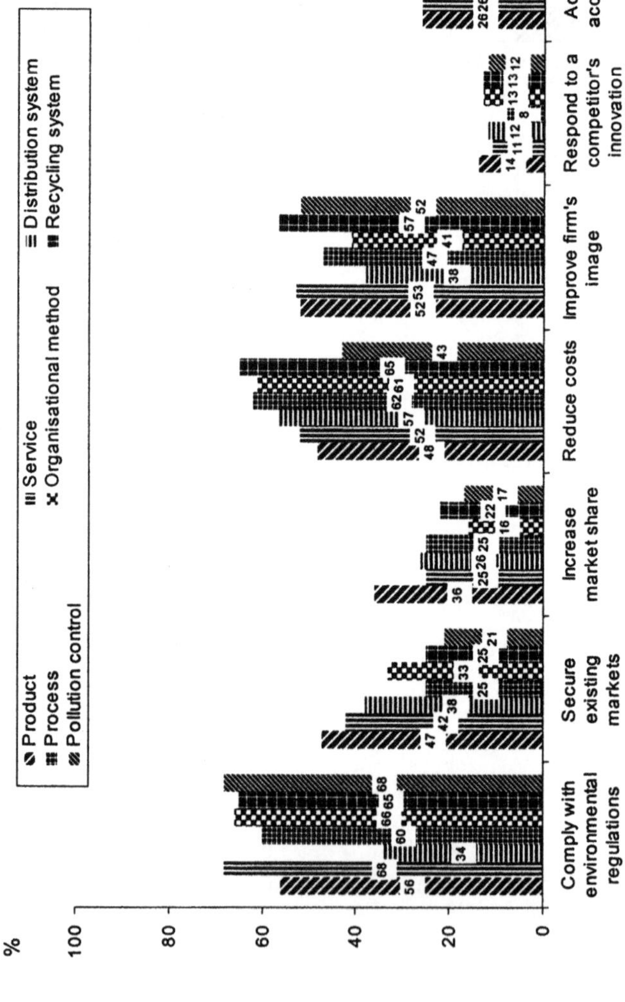

Figure 3: Effect of innovation on employment

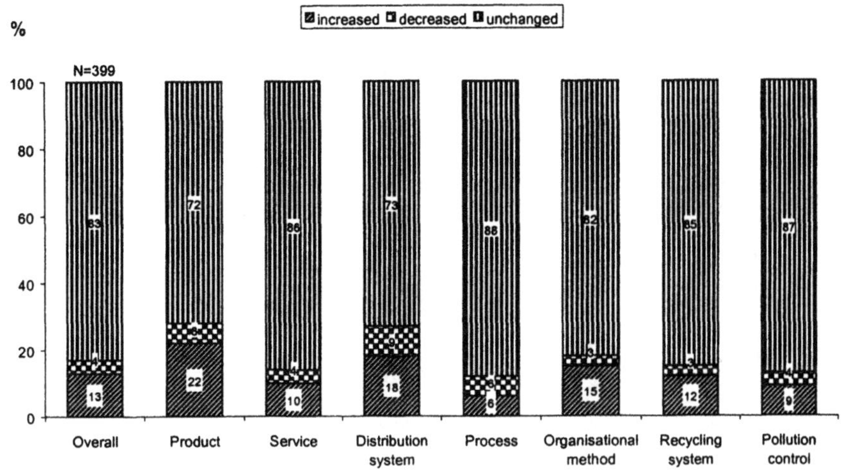

In 13% of the cases the number of long-term employees increased due to the innovation, in 4% of the cases it decreased. This result shows that there is a small but positive connection between the introduction of environmental innovations and employment.

Product innovations and innovations of the distribution system have a sizeable above-average positive employment effect (22% and 18%). Figure 4 shows by whom the eco-innovation was developed, overall and in connection with the question whether the innovation resulted in employment effects. For 37% of the firms, other firms or institutes developed the innovation (see left column). Against this, only 28% developed the innovation in their own institution.

Figure 5 gives an overview of the investment costs and their correlation with the employment effect of the eco-innovation. 46% of the firms invested less than € 50,000. The overall distribution of the share of investment costs is quite similar to the distribution of the investment costs for those firms which kept employment unchanged. This means that the level of investment costs is generally small.

Figure 4: Who developed the innovation

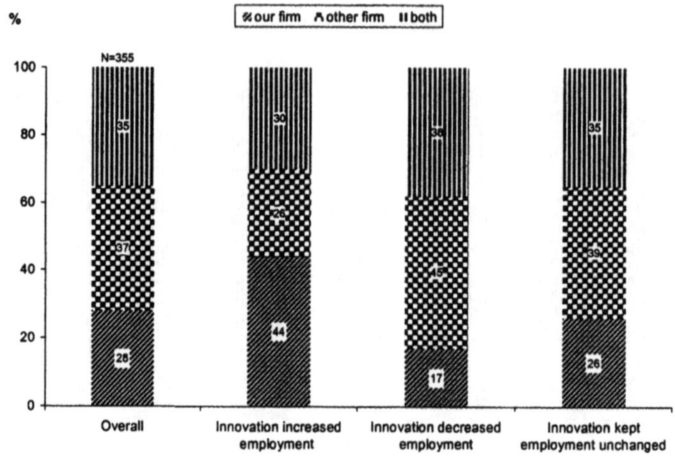

Figure 5: Investment costs of the innovation

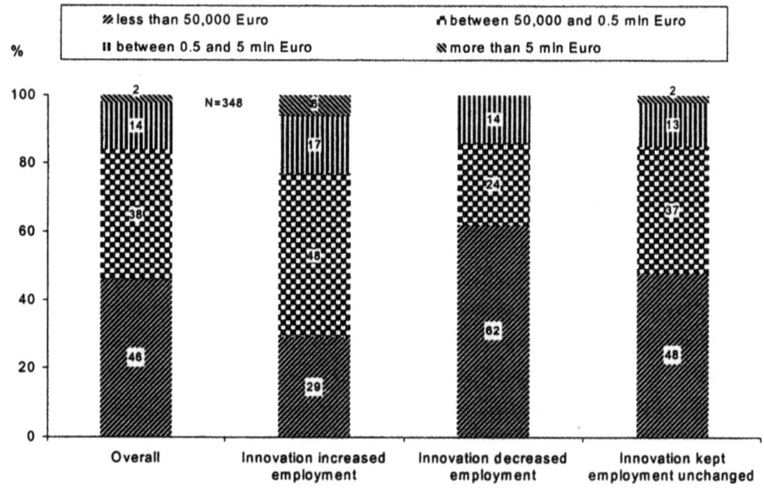

Figure 6 shows which proportions of the establishment's total innovation expenditures over the past three years were spent on the environmental innovation

Figure 6: Proportion of innovation expenditures

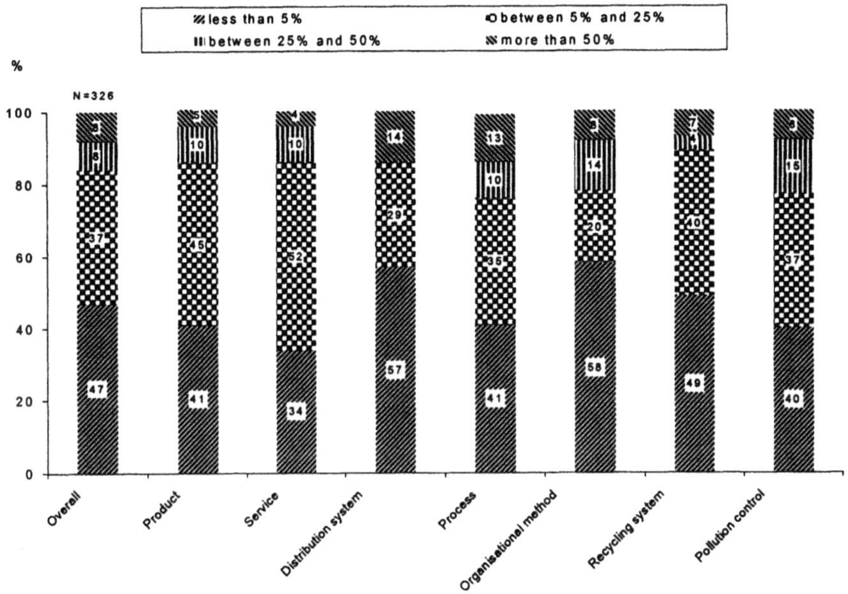

For the majority of establishments (47%), less than 5% were spent on the eco-innovation. Distributing the results by innovation types, it is remarkable that an above-average number of firms had a relatively higher investment share for product, service and process innovations, as well as end-of-pipe innovations. When comparing this result with Figure 3, there is no clear connection with the employment effects of such innovation types. Product and service innovations have a small, above-average positive effect, while process innovations have the same employment effect as distribution system innovations, the reference category. This means that there is no clear connection between the proportion of innovation expenditures and the employment effect.

Figure 7 gives an answer to the question whether there is a correlation between receiving subsidies or grants for introducing the innovation and the innovation type. Generally, only 8% of firms received a subsidy or grant. For service and pollution control innovations, an above-average number of firms received subsidies or grants. Only 1% of the establishments received subsidies.

An above-average share of innovators in the area of product and service innovations received subsidies. These are exactly the innovation categories which are labour-increasing compared to others. Thus it can be concluded that the allocation of subsidies for environmental innovation in Germany is employment-friendly.

Figure 7: Was subsidy/grant received (by innovation type)

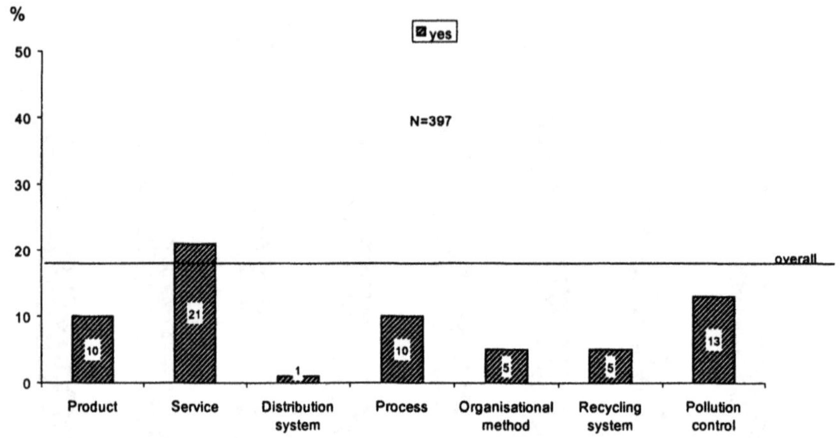

Figure 8 gives an answer to the question whether new skills required due to the eco-innovation are correlated with the innovation type and with the employment effect released by the same innovation. Overall, new skills were required by about 44% of the firms introducing environmental innovations. It can be shown that distribution system innovations (50%) and service innovations (48%) as well as product innovations (48%) imply a noticeable above-average need for new skills. Furthermore, there is a strong connection between the need for new skills and a positive employment effect: 69% of firms with new long-term employees due to the innovation realized a need for new skills.

Figure 8: Need for new skills by innovation type and employment effect

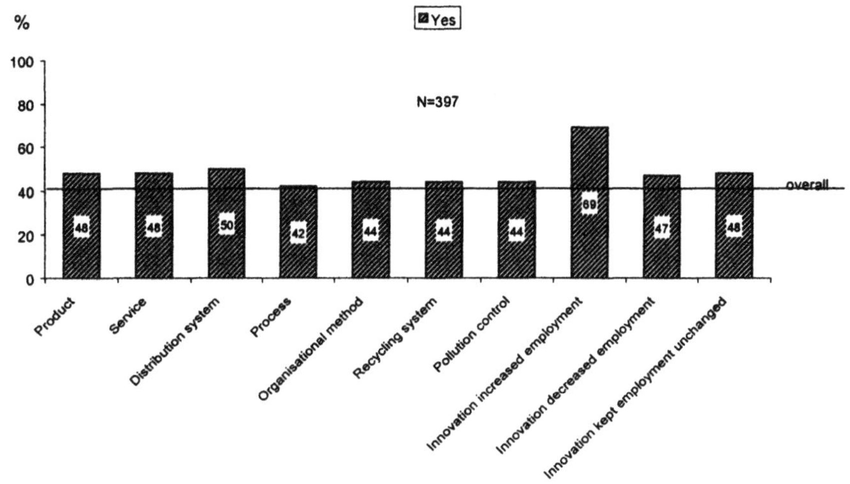

Figure 9 gives an overview of how the need for new skills was met (multiple answers were possible). The majority of the firms (90% of firms requiring new skills; 38% of all firms) uses the method of training existing employees.

Figure 9: How were new skills met

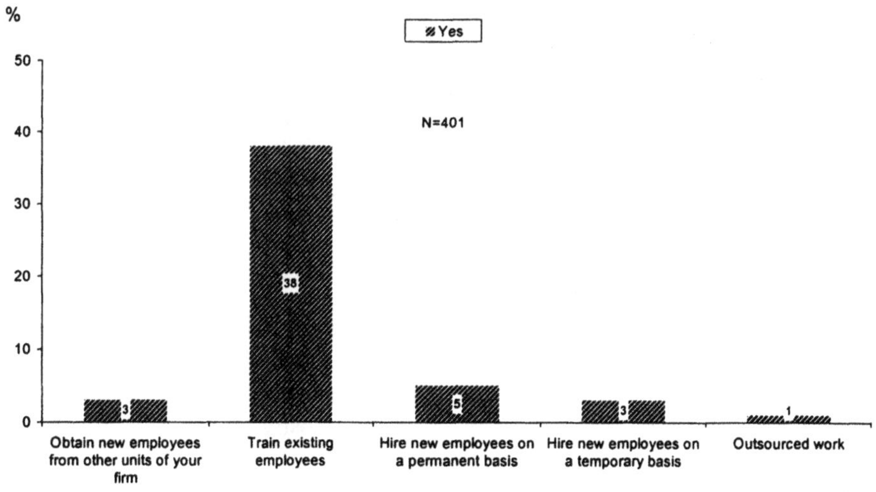

Only 8% of all firms (33% of firms requiring new skills) hire new employees on a permanent or temporary basis.

Figure 10 gives an answer to the question who received training and whether there is a connection with manufacturing or service firms or with the employment effect. Overall, mostly skilled workers received training (31%), followed by staff with college or university degrees (18%) and by unskilled workers (10%). This distribution is similar for both manufacturing and service firms. In firms which increased employment, employees received above-average training, especially the staff with college or university degrees. In firms which decreased employment, unskilled workers and staff with college or university degrees received training with a share of 11% and 8%. The generally high share of skilled workers can be explained by the innovations introduced. The firms obviously meet their demand for employees and skills for application but not for the development of the innovations, because they mostly have not developed the innovations themselves.

Figure 10: Who received training by industry and employment effect

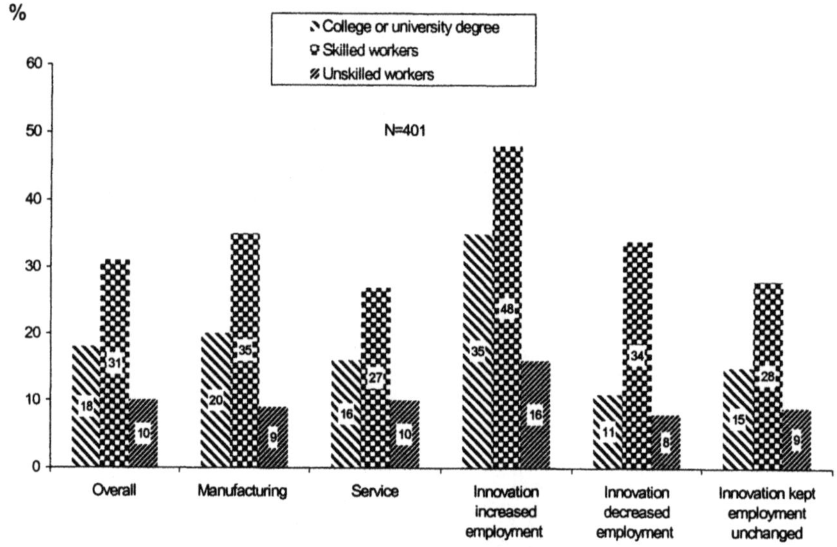

Figures 11a, 11b and 11c show the effects of the environmental innovations on sales, prices and costs. For more than 85% of the establishments, the innovations had no effect on sales and prices. For 15% of the establishments, sales even increased. Prices increased in 9% of the firms but mostly by less than 5%. Prices decreased in 12% of the firms (Figure 11a).

The energy costs decreased in more firms than the other costs forming part of this investigation. 37% of the firms could decrease their energy costs due to the innovation. This result goes in line with the high share of process innovations (see Figure 1) and the motive of cost reduction (see Figure 2). The level of decrease of energy costs also is relatively high: for 57% of the firms, energy costs decreased by more than 5%.

Figure 11a: Effects of innovations on sales and prices

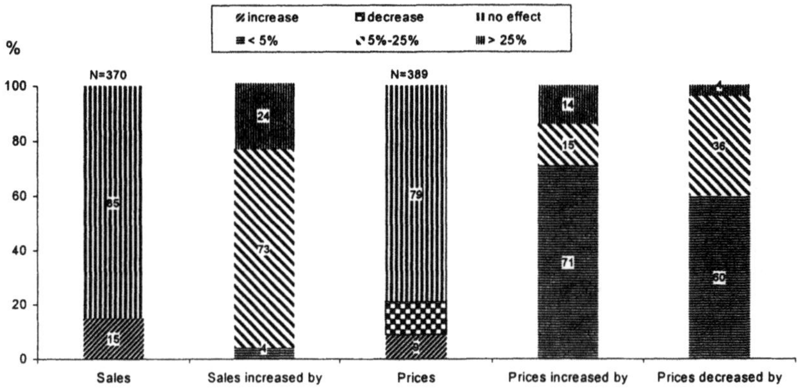

Material costs were unaffected for 58% of the firms. In those firms with changes of material costs, the number and level of decrease were a bit higher than in the firms with an increase (see Figure 11b).

Figure 11b: Effects of innovations on energy and material costs

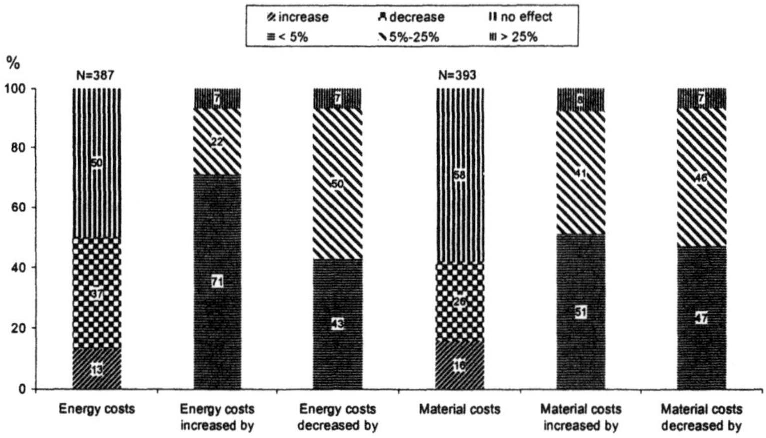

Waste disposal costs could be decreased by 43% of firms (see Figure 11c). This is due to the relatively high share of recycling innovations. The level of decrease was even more than 25% for 21% of the establishments. However, waste disposal costs increased for 17% of the firms.

Figure 11c: Effects of innovations on waste disposal and labour costs

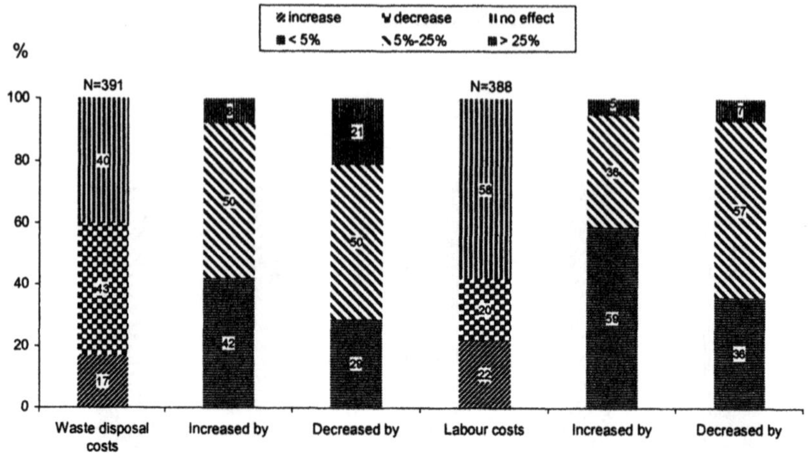

In 58% of the firms, labour costs did not change due to the innovation. However, they increased for 22% but mostly by less than 5%. Compared to the increased or decreased employment (see Figure 3), the higher percentage of firms with increased or decreased labour costs can be explained by the fact that labour costs increased due to the need for new skills. Another reason could be that in this survey, only the change in the number of long-term employees was taken into consideration. The number of short-term employees possibly changed, too.

3 General questions

Figure 12 gives an answer to the question how important environmental regulations are for processes and products and whether concrete changes were made in the last three years in order to comply with these regulations. It becomes apparent that environmental regulations seem to be important for both product and process innovations. Concrete changes in order to comply with the regulations were made in about half of the eco-innovative firms.

Figure 12: Importance of environmental regulations for processes and products and changes to comply with regulations

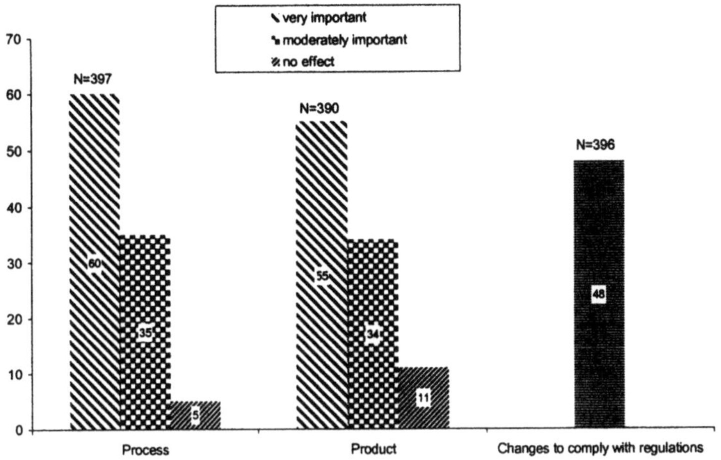

Price and quality can be clearly defined as the most important competition factors (see Figure 13). By a wide margin, they are followed by corporate image, environmentally friendly features and innovative products or services.

It is remarkable that environmentally friendly features are mentioned by only 5% of the firms as the most important factor, which is surprising because the answering firms are exclusively those which introduced environmental innovations in the last three years.

Figure 14 illustrates the changes in overall employment dividing the sample into large and small firms, manufacturing and service. The majority of establishments considered in the sample have recorded an increase in the overall employment (45% increase, 16% decrease, 39% unchanged).

Figure 14 furthermore shows that large and small firms increased employment almost to the same extent, and more service firms than manufacturing increased employment.

The right part of the figure provides an insight on the relationship between environmental innovation and related employment effect and, on the other hand, overall employment effect for the establishment considered. The results are very interesting: 79% of the establishments that reported an employment increase due to the eco-innovation also registered a general positive employment effect. The situation is completely reversed (and therefore consistent) for establishments where the eco-innovation determined a decrease in employment effect.

Figure 13: Important competition factors

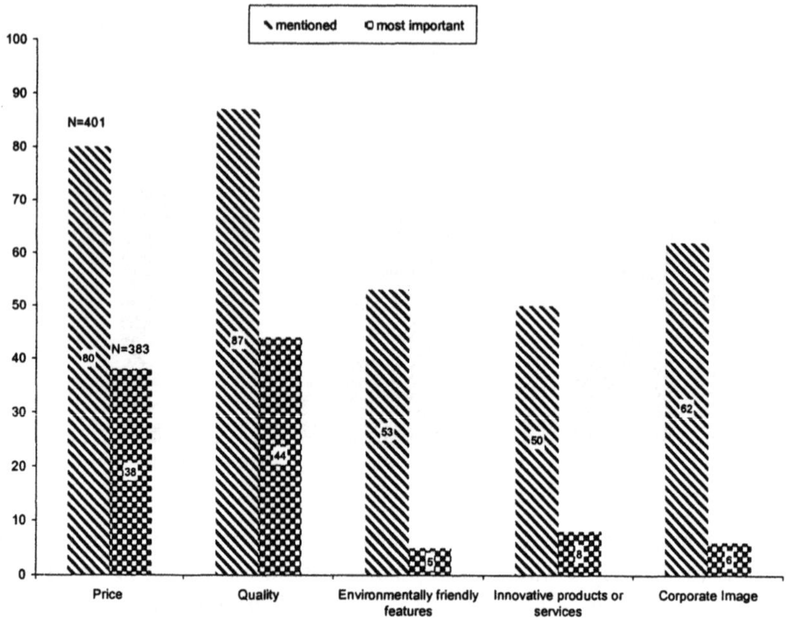

Figure 14: Change in overall employment

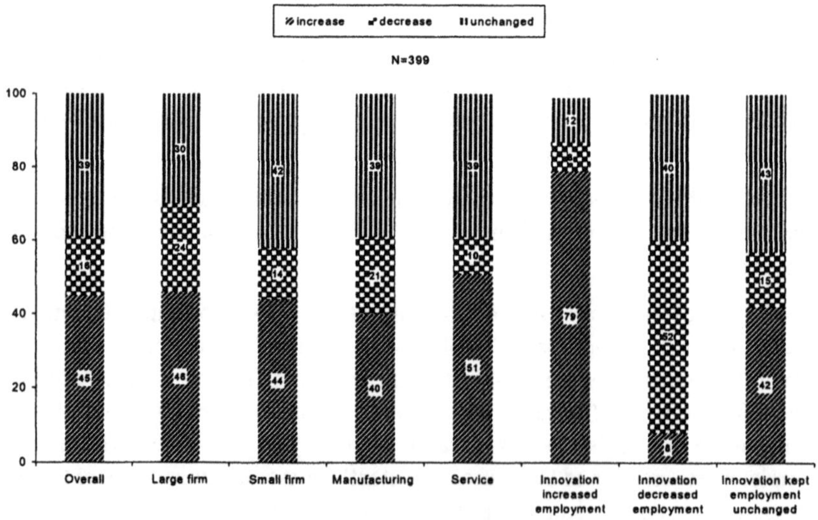

Figure 15 considers whether the establishment offers products or services that are specifically marketed as environmentally friendly and its relationship with the employment effect. The results demonstrate that 63% of companies that reported an increase in employment also sell products with tangible environmental characteristics.

Figure 15: Does your company offer products or services that are specifically marketed as environmentally friendly

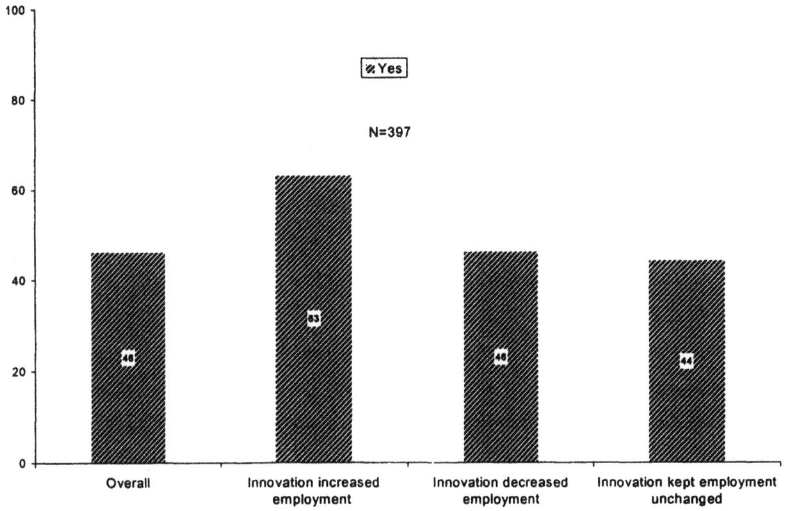

4 Conclusions

The German survey comes to the result that in general environmentally beneficial innovations have no sizeable effect on employment. However, there are more firms that increased rather than decreased their number of employees in the wake of the eco-innovation. In addition, it can be concluded that the positive employment effect, entailing increasing labour costs, had no negative impact on sales even if prices often had to be raised. The positive influence on the image of the firm and the quality of the products seems to compensate for the higher prices.

In addition, a further positive employment effect can be expected from the tasks outsourced due to the eco-innovation. The effect cannot be quantified in this study. Also other indirect employment effects like those resulting from the firms that had developed the eco-innovations cannot be quantified either.

The three most cited reasons for introducing the innovation are to improve the firm's image, to comply with environmental regulation and to reduce costs. Market goals play only a minor role.

Besides the employment aspects, many firms were able to decrease prices because other important costs like energy costs and waste disposal costs declined considerably due to the eco-innovation.

An above-average share of innovators in the area of product and service innovations received subsidies. These are exactly the innovation categories which are labour-increasing compared to others. Thus it can be concluded that the allocation of subsidies for environmental innovations in Germany is employment-friendly. However, there is still some potential for further shifts towards a support of cleaner production technologies.

The Dutch survey

Anthony Arundel, Leann Chervenic-Poeth and René Kemp

1 Aims and methodology

The survey seeks to determine the employment and skills consequences of seven categories of environmental innovation for the innovating company and how such effects are related to factors such as the motivation for innovating (cost reduction, compliance with environmental regulations and so on) and to company and market characteristics (the basis for competition).

A random sample of 253 firms active in manufacturing and 420 in selected service sectors in the Netherlands was drawn. Each firm was contacted in order to identify a suitable respondent to the survey, who was usually the R&D manager or the CEO for small firms. The manufacturing sector included NACE sectors from NACE 15 to NACE 36 inclusive. In addition, firms active in utilities (electricity, gas and water) and in construction were sampled. The service sector sample was limited to five sectors: 1) wholesale, retail and trade; 2) hotels and restaurants, 3) transport, storage and communication; 4) finance, and 5) real estate, renting and business activity. Excluded sectors include, among others, agriculture, public sectors and health care.

299 firms agreed to participate. This is equivalent with a response rate of 44.4%. Of the 299 participating firms, 208 (69.6%) had introduced one or more environmental innovations and agreed to proceed to the other survey questions. The number of firms contacted, the number of responses, and the response rates are given in Table 1.

The response rate is higher in the manufacturing/industrial sectors than in services, possibly because the latter were less likely to find the questionnaire topic of relevance to them. The response rate was also higher for large (57.8%) than for small (40.2%) firms.

All analyses are limited to a maximum of 208 environmentally innovative firms that stated, during the interview, that they had introduced one or more innovations with an environmental benefit in the three years preceding the interview. These firms are called eco-innovators. In order to increase the cell counts to a reasonable number for analysis, all manufacturing and other industrial sectors were combined into one group of 117 innovative firms, while all service sectors were combined into a second group of 91 innovative firms. Of the 208 eco-innovators, 158 firms

(76%) have less than 200 employees, while 50 (24%) have 200 or more employees.

Table 1: Response rate for the Dutch IMPRESS survey

Sector category	Total sample	Total responses	Response rate	Responses from innovators
Large services	88	46	52.3%	28 (60.9%)
Small services	332	131	39.5%	85 (64.9%)
Large manufacturing	73	47	64.4%	41 (87.2%)
Small manufacturing	180	75	41.7%	54 (72.0%)
Total	673	299	44.4%	208 (69.6%)
Small firms	512	206	40.2%	
Large firms	161	93	57.8%	

Table 2 shows the distribution of the 299 responding firms by sector.

Table 2: Responses by sector

Sector	Number of firms	%
D – Manufacturing	113	37.8
E – Electricity, gas and water	3	1.0
F – Construction	30	10.0
G – Wholesale, retail and trade	52	17.4
H – Hotels and restaurants	8	2.7
I – Transport, storage & communication	14	4.7
J – Financial intermediation	37	12.4
K – Real estate, renting and business activities	42	14.0
TOTAL	299	100%

1.1 Mean values for employment, sales and investment costs

The survey questions on the cost of the firm's most beneficial eco-innovation, 1999 employment, 1999 sales, and 1999 labour costs as a percentage of sales provide two response options for the respondents. The first option asks for a point estimate, for example a respondent can answer that the innovation cost 0.9 mln €. The second option provides several response categories. The final category is always open ended. For analysis, the point and categorical response categories are combined after providing a best point estimate for the categorical responses. This best estimate is based on the mean value for firms that provided a point estimate within the relevant categorical boundaries. For example, the mean innovation costs for firms that gave a point estimate of over 5 mln € is 6.5 mln. This value is assigned to all firms that gave a categorical response of 'over 5 mln €'.

1.2 Description of the environmental innovators

The *manufacturing firms in the sample are slightly larger on average than the service firms*, with 28% of manufacturing and 19% of service firms having 200 or more employees in 1999. Table 3 gives data on sales and employment in the two main sectors.

There is no difference in the mean sales of manufacturing and service firms, but the former are larger, with an average of 290 employees compared to 213 for the service firms. However, both sales and employment have increased more among service than among the manufacturing firms. For example, 66% of manufacturing firms reported an increase in sales over the previous three years, compared to 83% of service firms. Over this time period average employment declined among the manufacturing firms by 7.7%, compared to an increase of 27.6% among the service firms. Summed across all firms in each sector, the *manufacturing firms lost an estimated 876 employees while the service firms gained 2,375 employees.*

There is no difference in the percentage of employees that have a tertiary degree between manufacturing (14%) and service (17%) firms, but a significantly higher percentage of employees among large firms have a tertiary degree compared to small firms (22% vs. 13%, $p = .003$). A second question asked if the number of employees with a tertiary degree had increased or decreased by more than 5%. Neither the difference by sector (36% of manufacturing firms vs. 39% of service firms report an increase of more than 5%) or firm size is statistically significant.

Table 3: Sales and employment data for manufacturing and service firms

	Change over previous 3 yrs				Mean change[1]	p[2]
	Inc.	None	Decr.			
Manufacturing sales[3]	66%	7%	27%	100%	8.7%	
Service sales[3]	83%	4%	13%	100%	14.9%	.03
Manufacturing employees	43%	18%	39%	100%	-7.7%	
Service employees	66%	8%	26%	100%	27.6%	.05

[1] Respondents were asked to give a point estimate of the percentage change in the firm's sales and employment. Respondents that were unable to provide a point estimate were given a choice of several categories. The mean value for each category was determined from the averages for firms that gave point estimates within each category.

[2]Significance of the difference in the mean change in sales or employment for manufacturing and service firms.

[3]mln €.

2 Types of environmental innovation introduced

All respondents were asked if they had introduced one of the following types of innovations with environmental benefits *within the previous three years*:

1. Pollution control techniques
2. Recycling systems
3. Product change (including service innovations)
4. Process changes
5. Organisational methods
6. Distribution systems.

A firm could reply to only one of the six choices or to all six of them. In these analyses, firms that responded 'Don't know' are assumed not to have introduced the relevant innovation. The results are given in Table 4.

Table 4: Types of environmental innovation introduced by 208 Dutch eco-innovative firms

	% introducing this innovation	% only introducing this innovation	Mean no of 6 environmental innovations introduced by these firms
Organisational	57% (119)	2.9% (6)	4.0
Pollution control	52% (109)	0.5% (1)	4.1
Recycling	56% (116)	1.4% (3)	4.0
Product	53% (111)	2.9% (6)	3.9
Process	71% (148)	1.4% (3)	3.9
Distribution	44% (91)	2.4% (5)	4.1
	100% (208)	12% (24)	

The *most frequently cited environmental innovation is process innovation*, cited by 71% of the firms, followed by product innovation (53%) and pollution control (52%). The lowest share is for distribution systems (44%), an innovation that is usually not included in eco-innovation surveys, but that is still quite important.

Only 24 firms introduced just one of the six possible environmental innovations and only 2.9% of the 208 firms just introduced an organisational innovation. Not surprisingly, larger firms introduced a significantly greater number of different types of eco-innovations compared to small firms (average of 3.8 vs. 3.2, $p = .02$). Manufacturing firms also introduced more eco-innovations than service firms (average of 3.5 vs. 3.1, $p = .07$).

3 Most important environmentally beneficial innovation

The respondents were asked a series of questions on the firm's most *environmentally beneficial innovation*, for which there are 206 valid responses. This innovation could be a single type (i.e. product or organisational) or combine up to eight categories: a product, service, distribution system, process, organisational method, recycling system, pollution control, or 'other' type of innovation. *Only 43 (21%) of the most beneficial innovations were in only one category.* Only one firm

(0.5%) cited an organisational innovation alone. Figure 1 gives the distribution by sector for each environmental innovation.

Figure 1: Sector distribution of the most environmentally beneficial innovation types

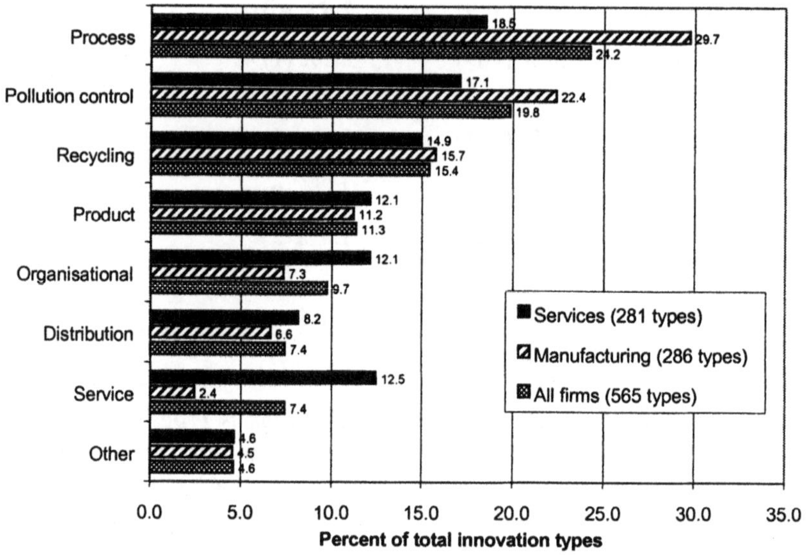

The sector totals in Figure 1 sum to 100%. For example, the 117 manufacturing firms cite 286 separate innovations, of which 85 (or 29.7%) are process innovations. In manufacturing, 67.8% of all cited innovations are due to three types: processes, pollution control, and recycling, compared to only 50.5% for service firms. Figure 1 clearly shows that manufacturing firms are more likely to cite process and pollution control innovations while service firms are more likely to cite organisational and service innovations. There are only small differences by sector for recycling, product, distribution and 'other' environmental innovations.

Table 5 gives the percentage of firms that cited each of two innovation categories for their most environmentally beneficial innovation. For example, 8.3% of the 206 firms cited both 'service' and 'product' innovation. Recycling, pollution control and process innovations frequently occur together. In fact, 85.4% of the 206 innovations include one or more of these three categories, while 20% include all three.

Table 5: Percent of the most environmentally beneficial innovation that fits two categories

	Service	Distrib.	Process	Org.	Recycle	Poll. contr.	Other
Product	8.3	6.8	19.4	6.8	11.2	15.0	3.3
Service		6.8	11.7	11.2	9.2	9.2	1.6
Distribution			9.7	10.7	9.7	9.7	0.0
Process				16.5	30.6	42.2	5.8
Organisation					15.5	15.0	2.9
Recycling						25.2	3.9
Pollution control							5.3

Organisational innovations are also more likely to be linked with these three types of innovations than with product, service, distribution, or 'other' innovations. In fact, 48 (87.3%) of the 55 innovations that include an organisational component are linked with one or more of the process, recycling, or pollution control categories. This is somewhat surprising, since a relatively high percentage of the organisational innovators are in the service sector, as shown in Table 6.

Table 6: Distribution of organisational and non-organisational innovators

Org. innovation	Small manufacturing	Large manufacturing	Small services	Large services	
No (151)	47.7%	15.2%	31.8%	5.3%	100%
Yes (55)	20.0%	18.2%	45.5%	16.4%	100%

Notes: small = 10 to 200 employees, large = 200+ employees. The difference in distributions between organisational and non-organisational innovators is statistically significant, with $p < .001$.

Respondents were asked a number of questions about the innovation which they considered to be the most important from an environmental point of view. These questions concerned the employment and skills effects and costs effects of adopting the innovation, among other things.

The fact that the question asking the respondent to describe the most environmentally beneficial innovation includes up to eight different categories creates a problem: it is very difficult to link a particular type of innovation to a specific outcome, such as a change in employment or costs. Therefore only descriptive results are provided for each type of eco-innovation. The results given below by innovation type are almost always for an eco-innovation that spans several categories. In this respect, the results are for innovations that contain a *component* of each category, but which can also include other categories.

4 Proportion of innovation expenditures

Respondents were asked to estimate their total investment costs for their most beneficial eco-innovation. The estimated average cost of each type of eco-innovation, in €, is given in Table 7[1].

The most expensive types of innovations for service firms contain a product or service component, while product and service eco-innovations are much less costly for manufacturing firms than eco-innovations that contain a process, pollution control, or organisational component. Overall, manufacturing firms spent almost five times as much on the eco-innovation than service firms.

Figure 2 gives the mean expenditure shares for the firm's most beneficial eco-innovation. It shows that eco-innovations with a process component account for a high percentage of total innovation expenditures: for 34% of the companies the process innovation costs less than 5% of total innovation expenditures, while for 19% it was more than 50% of total innovation expenditures. New distribution systems consumed an even higher share of total innovation expenditures, while the lowest shares of total innovation expenditures were for organisational and recycling innovations. For all firms combined, the most beneficial eco-innovation accounts for more than 50% of total innovation expenditures in 18% of the cases, which suggest that eco-innovation for these firms is probably one of their most important innovative activities. For 61% of the companies, the eco-innovation amounted to more than 5% of the innovation expenditures, and for 36% more than 25%. It shows that firms spend a comparatively large fraction of their total innovation investment on innovations with an environmental benefit, although for some of these firms the environmental gain can be unintended.

---

1 In total, 150 firms gave a point estimate, 42 gave a category estimate, and 16 did not answer either version of the question. 33 of the firms that gave a point estimate gave the cost as zero. There was no difference in 1) who developed the innovation or by 2) the type of innovation between firms that gave zero and positive costs. The boundaries for the category options (in Euro equivalents) are less than 50,000 €, between 50,000 € and 0.5 mln €, between 0.5 mln € and 5 mln € and more than 5 mln €.

Table 7: Mean investments costs ('000 €) for the most environmentally beneficial innovation by innovation type

	Manufacturing	Services	All firms
Product	244.9	1,163.7	736.0
Service	51.4	1,059.3	904.3
Distribution	259.9	744.7	553.2
Process	4,061.7	281.8	2,585.1
Organisational	3,534.4	657.9	1,764.3
Recycling	1,831.9	279.8	1,055.9
Pollution control	5,187.3	495.8	3,126.6
Other	147.2	313.5	233.7
All types	3,043.8	602.9	1,937.7

Figure 2: Distribution of innovation expenditures

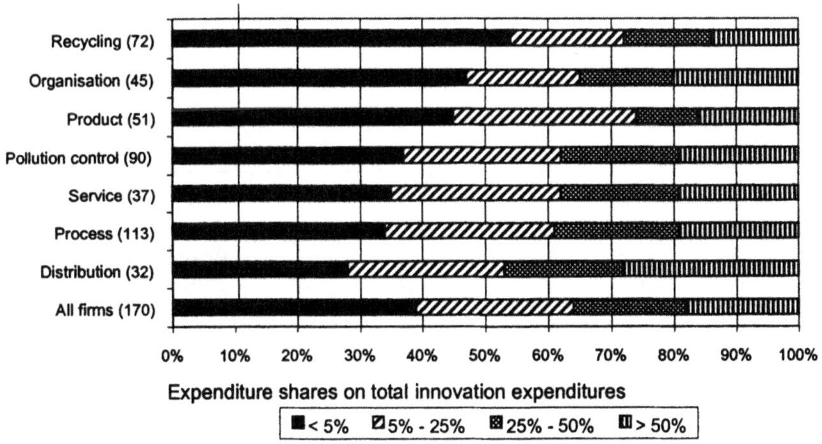

5 Subsidy/grants received for the innovation

Overall, 22% of the firms received a subsidy or grant for their most beneficial eco-innovation, with the highest subsidy rate among large manufacturing firms (27%) and the lowest rate for small service firms (20%). The percentage of firms that received a subsidy by the type of eco-innovation is reported in Figure 3. The highest subsidy rate is 29% for eco-innovations with a product component, which is much higher than the rate for Switzerland of 7%. The proportion of eco-innovations receiving a subsidy or grant is also high for the other types of eco-innovations: 23% for process innovations, 22% for service innovations, 21% for pollution control innovations and respectively 15%, 16% and 17% for distribution systems, organisational methods and recycling systems. Apparently, the Netherlands has plentiful opportunities for getting innovation subsidies, and companies know how to obtain one. Unfortunately, we cannot tell from the data if subsidy rates are higher for intentional environmental innovations compared to innovations with unintended environmental benefits.

Figure 3: Was subsidy/grant received (by innovation type)

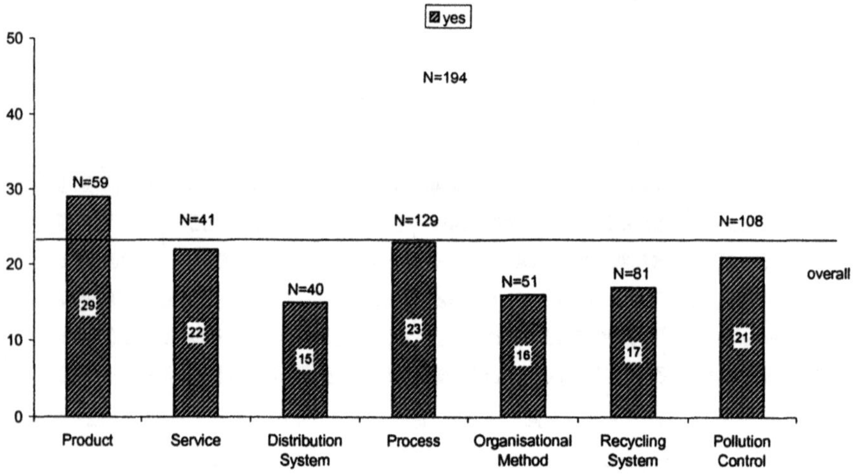

6 Important reasons for introducing the innovation

What motivated the companies to adopt or develop the innovation? Do these motivations differ across innovation types? The percentages of respondents that cite at least one of seven reasons by type of eco-innovation are given in Figure 4. The seven reasons are:

1. Comply with environmental regulations (compliance)
2. Secure existing markets
3. Increase market share
4. Reduce costs
5. Improve firm's image
6. Respond to a competitor's innovation
7. Achieve accreditation.

The most often stated important reason for introducing the innovation is to improve the firm's image, followed by reducing costs. Responding to a competitor's innovation is least frequently mentioned. Unfortunately, we do not have information about the *single* most important reason for introducing this eco-innovation. The fact that improving the firm's image is the most frequently cited does not mean that it is the most important reason. In fact, it is unlikely to be the most important reason.

According to other surveys (Williams et al., 1991, Garrod and Chadwick, 1996, Green et al., 1994), the most important reason for adopting an environmental technology is compliance with environmental regulations, with cost reduction usually ranking second. This finding is neither confirmed nor refuted by our survey. Our survey simply shows that there are many different important reasons – besides complying with regulations – for introducing an eco-innovation. These are: improving firm's image, reducing costs, achieving an accreditation, and, for product and service innovations, securing existing markets and increasing market share.

Compliance with environmental regulations is more frequently cited for pollution control (mentioned by 68% of the respondents), than for the other types of eco-innovation, especially service, distribution, and product innovations (mentioned by respectively 45%, 51% and 56% of the respondents). For process innovations and recycling, compliance with environmental regulations is mentioned by 64% and 62% as an important factor, a share almost as high as the share for pollution control (68%). It thus shows that these types of innovations have an important compliance factor but also shows that there are other important reasons for introducing such innovations, such as cost savings (not environment-related) and improved image (environment-related).

Figure 4: Important reasons for introducing the innovation

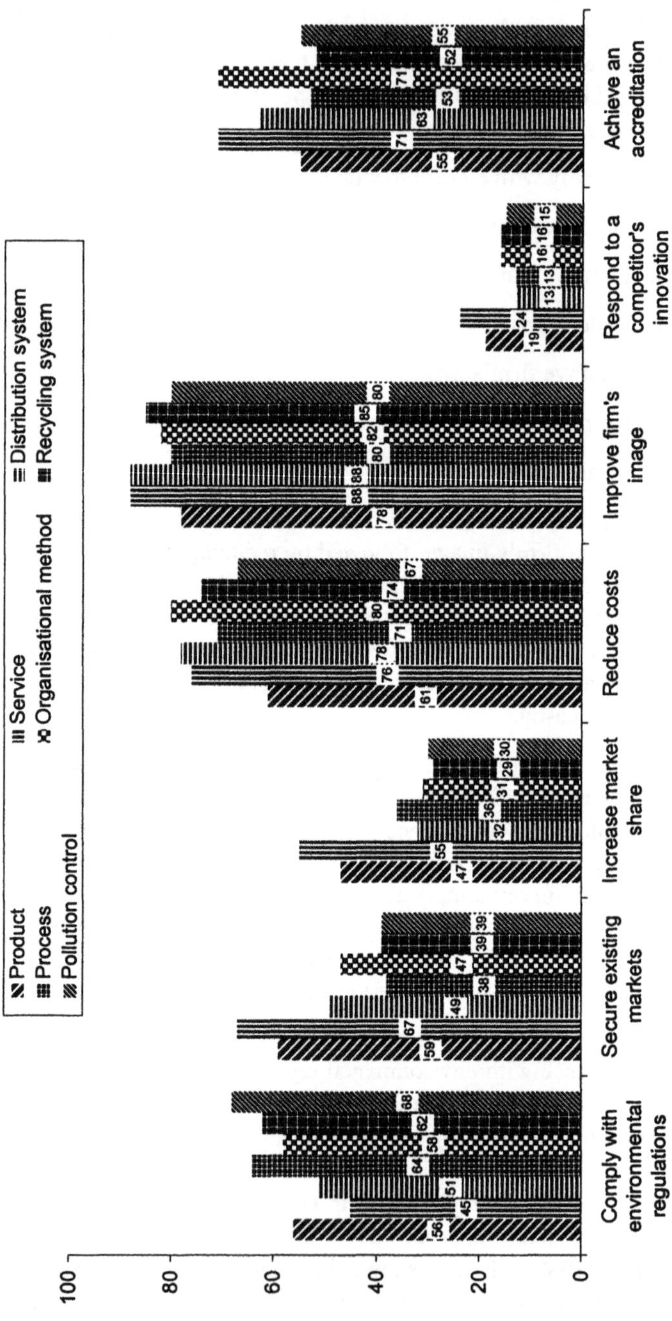

Another important factor, besides the ones noted, is achieving an environmental accreditation, ranking third for service innovations, distribution systems and organisational methods. 71% of the respondents mentioned it is an important reason for introducing a service innovation and organisational innovation.

On average, 3.4 reasons were cited by 205 respondents as being important for adopting/developing their most important environmental innovation. Only 24 firms cited just one reason, while another 5 stated that none of them were a factor. There are no significant differences in the average number of reasons cited by manufacturing firms (3.3) and services firms (3.5) or by firm size.

Is there a relationship between the motivation for innovation and the economic performance of the company? To analyse this, we examined the correlation between innovation motivation and the effect of the eco-innovation on total sales. The results of this analysis are given in Table 8.

Table 8: Percent of firms citing an increase in sales by the reason for introducing the most environmentally beneficial innovation

Reason for introduction	N[1]	Reason cited	Reason not cited	P[2]
Compliance	120	15.0%	22.4%	.13
Secure existing markets	77	31.2%	9.2%	.00
Increase market share	56	46.4%	6.4%	.00
Reduce costs	126	20.6%	12.9%	.12
Improve firm's image	153	19.0%	14.0%	.31
Respond to competitor's innovation	23	43.5%	14.5%	.00
Achieve accreditation	101	22.8%	12.6%	.05

[1] Number of firms citing the reason. [2] Significance level of difference based on chi-square comparison with firms that do not cite the reason.

None of the firms reported a decrease in total sales, so Table 8 only gives the percentage citing an increase. For example, 15% of firms that cite compliance report an increase in total sales, compared to 22.4% that do not cite compliance. In this case, there is no significant difference, with $p = 0.13$. Of interest, 'compliance' is *the only reason for which total sales increased more among respondents that did not cite this reason.*

Firms that cite 'securing markets', 'increasing market share', 'respond to competitor' and 'achieve accreditation' are more likely than firms that do not cite these reasons to report an increase in sales. However, it is not clear which are the driving factors for firms to cite more than one reason for introducing their most beneficial eco-innovation.

7 Employment effects of types of innovation

Respondents were asked about the employment effects of their most beneficial environmental innovation – whether it increased employment, decreased employment or did not change employment in the innovating firm. The results are given in Figure 5.

Figure 5: Effect of innovation on employment

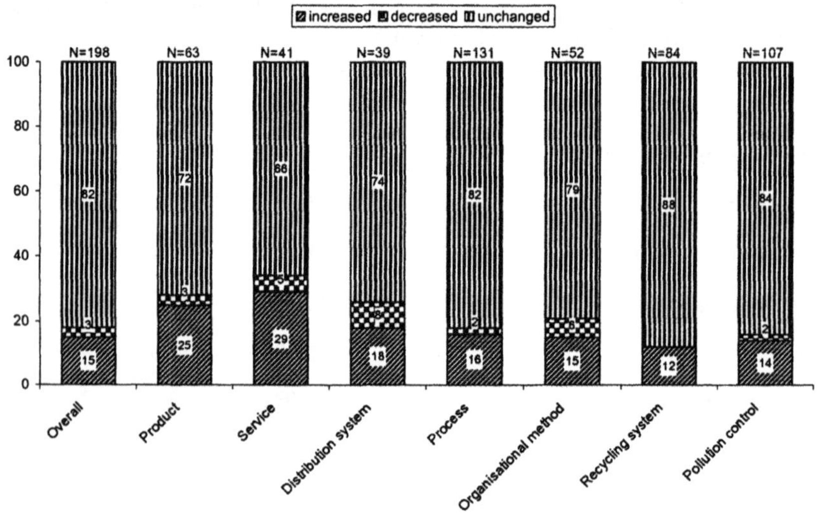

The nature of the employment effect (whether it led to an increase or decrease of employment in the innovating company or had no effect) is given for different types of eco-innovation. For the vast majority of the firms, 82%, the eco-innovation did not change employment. In the instances in which employment was changed, employment was most often increased (15% of the eco-innovating companies experienced an *increase* in employment due to the introduction of the innovation; 3% experienced a decrease). The *highest employment increases were for service and product innovations,* where the introduction of the innovation led to an increase of employment in 29% and 25% of the companies (against 5% and 3% of the companies experiencing a decrease in employment). In the case of innovations with a process, pollution control and recycling component, the number of companies experiencing an increase in employment was 16%, 14% and 12% respectively.

One possible explanation for the relatively positive employment effect for product and service innovation is that these innovations could more likely be developed in-house (creating employment in innovation, marketing and research) than many environmental process innovations and pollution control techniques, which could be purchased from capital equipment suppliers. Table 9 provides the percentage of each type of eco-innovation that was developed in-house, by other firms or organisations ('external'), by a combination of in-house and external efforts ('both'), and in an 'other' manner. As expected, innovations with a process, recycling or pollution control component are less likely than innovations with a product or service component to be developed in-house. For example, 40.5% of innovations with a service component were developed in-house and only 16.7% of them were developed externally. In comparison, 25.2% of innovations with a pollution control component were developed in-house and 36% were developed externally.

Table 9: Developer of eco-innovations by type of innovation

	In-house	External	Both	Other	
Product	37.5%	25.0%	32.8%	4.7%	100%
Service	40.5%	16.7%	42.9%	0.0%	100%
Distribution	45.2%	23.8%	28.6%	2.4%	100%
Process	29.6%	35.6%	31.1%	3.7%	100%
Organisational	34.5%	23.6%	41.8%	0.0%	100%
Recycling	29.1%	29.1%	39.5%	2.3%	100%
Pollution control	25.2%	36.0%	33.3%	5.4%	100%
Other	11.5%	50.0%	19.2%	19.2%	100%

The positive employment effect from innovations with an organisational component, however, is less than expected: as shown in Figure 9, only 15% of these firms experienced a positive employment effect and 6% report a decline in employment. We expected the employment effect for organisational innovation to be above the average of 15%, as the introduction of, for example, an environmental management and auditing system should bring additional work in the form of monitoring and reporting. Additional analysis of the type of organisational innovation is needed to understand this.

Figure 6 confirms that the *employment effects are more positive for innovations developed by the company itself.* In 41% of the cases where the innovation increased employment, the innovation was developed by the company, compared to 32% of all cases. In 75% of the cases where the innovation decreased employment, the innovation was developed by another firm, although this result is based on only four firms.

Figure 6: Who developed the innovation?

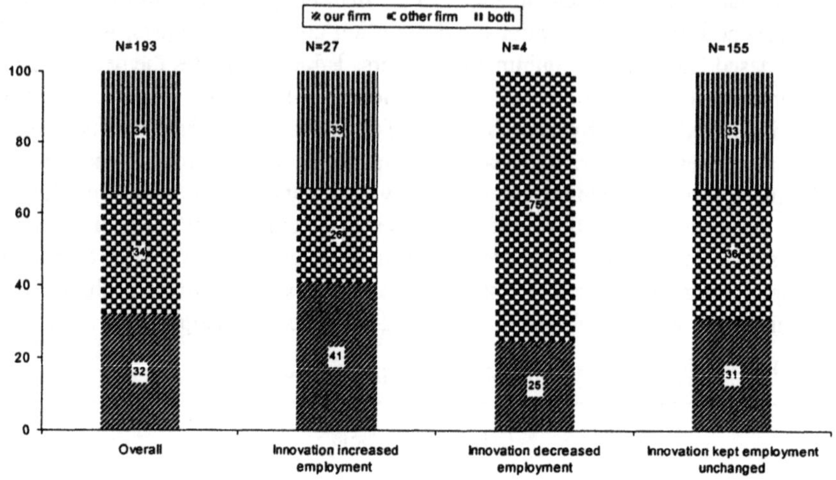

Are innovations that increase employment more expensive than those that have no effect or which reduce employment? Figure 7 provides some information on this.

Figure 7: Investment costs of the innovation

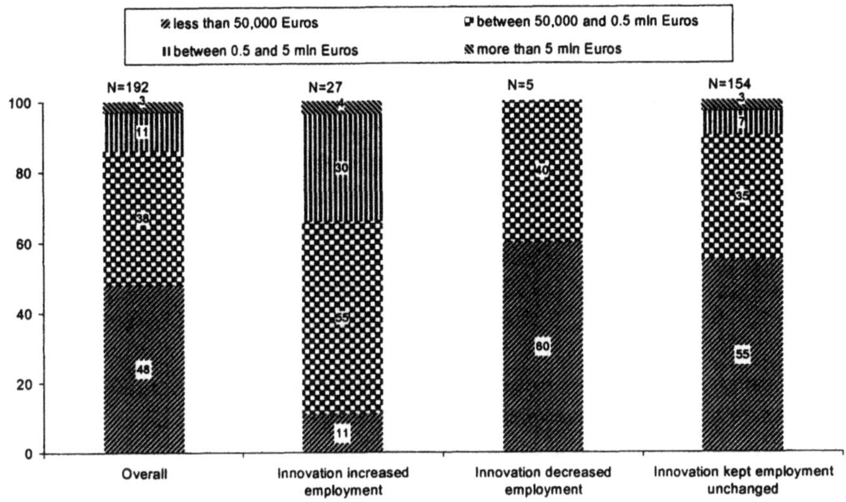

Figure 7 shows that *innovations that increase employment (in the innovating company) are relatively expensive compared to innovations with no effect on employment*: 89% of the innovations that increase employment in the innovating company cost more than 50,000 €, 34% cost more than 500,000 € and 4% are above 5 mln € (excluding firms that responded 'don't know' to innovation costs). In comparison, 55% of the innovations that have no effect on employment cost less than 50,000 €. (The difference in the categorical cost of innovations that increase and have no effect on employment is statistically significant, with $p < .000$.) This is an interesting result, and something we did not know from other studies. An even higher percentage of innovations that decrease employment cost less than 50,000 €, but this is based on only five firms.

8 The skills consequences of the innovation

The skills implications of the eco-innovation are depicted in Figure 8. The introduction of the most important eco-innovation required new or different skills for 32% of the firms, with the largest share for innovations with a product (44%), process (39%), pollution control (38%) and service (37%) component. It is unclear how these figures compare to normal innovations, but the skill requirements appear quite notable. Of course, the results are for the most beneficial eco-innovation, which could tend to be major innovations. For innovations that increased employment, new or different skills were required by 59% of the firms, compared to only 25% of the firms where the innovation had no effect on employment.

Figure 8: Need for new skills by innovation type and employment effect

Another question is how the demand for new skills was met. The answer is given in Figure 9, which shows that the *most important method for meeting the skills requirements of the eco-innovation is training existing employees*. Outsourcing ranks second, with 56% of the companies doing this, followed by hiring new employees on a permanent basis. Hiring new employees from other units of the company ranks lowest with 21%. This option is only open for larger companies. Hiring people on a temporary basis is done by 31% of the companies. For less important innovations the hiring of people outside the company and outsourcing are likely to be less common methods for meeting new skill requirements.

Figure 9: How were new skills met?

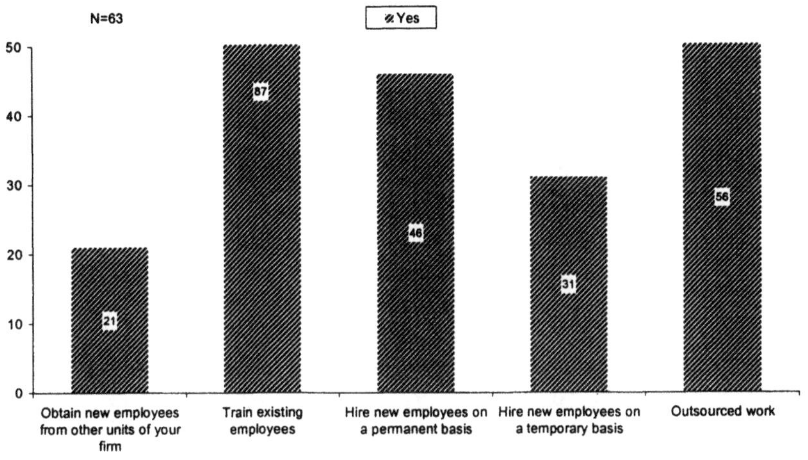

9 Effects of the innovation on sales, prices and costs

Respondents were asked about the effects of the innovation on their sales, prices and costs. We asked them whether the innovation led to an increase or a decrease in sales or whether it had no effect. The same was asked for prices. The results are given in Figure 10.

Interestingly, the adoption of the most environmentally beneficial innovation did not have a negative effect on sales. None of the companies experienced a reduction in sales as a result of adopting the innovation. For 18% of the companies it increased sales (by less than 5% for 23% of the companies, 5-25% for 56% of the companies and more than 25% for 21% of the companies experiencing a sales increase). For prices the results are mixed: the innovation increased prices for 12% of the companies and reduced prices for 14%. Most of the price increases were in the range of 5-25%.

The effects of the innovation on energy costs and on material costs are given in Figure 11. It shows that the innovation helped to decrease energy costs for 41% of the companies and reduced material costs for 34%. It increased energy costs for 11% of the companies and increased material costs for 22%. The decreases in costs exceeded the increases in costs. 65% of the increases in energy costs were below 5%, whereas 61% of the decreases were above 5%. For materials, 51% of the cost increases were below 5% and 62% of the decreases above 5%. The cost decrease effect dominates thus over the cost increase effect for these two cost categories: energy and materials. This corresponds with the often stated assertion that eco-innovations often reduce energy and material costs.

Figure 10: Effects of innovations on sales and prices

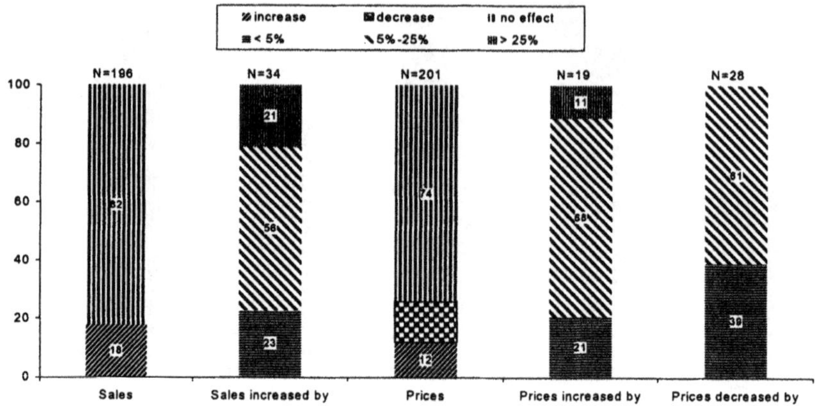

Figure 11: Effects of innovations on energy and material costs

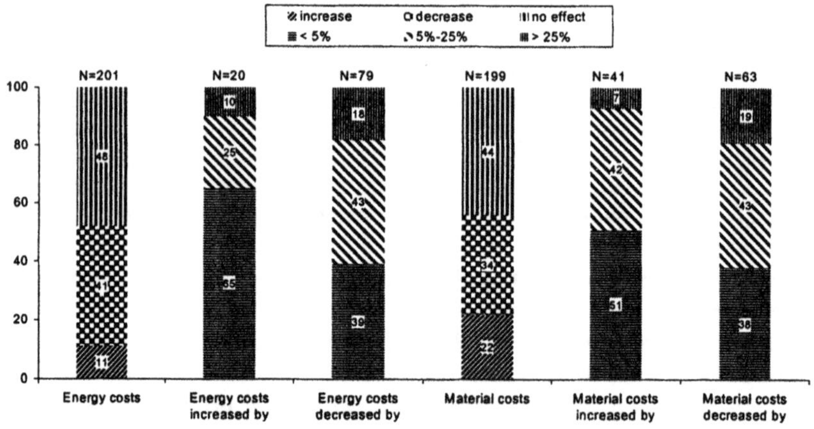

The effects of the innovation on waste disposal costs and on labour costs are given in Figure 12.

Figure 12: Effects of innovation on waste disposal and labour costs

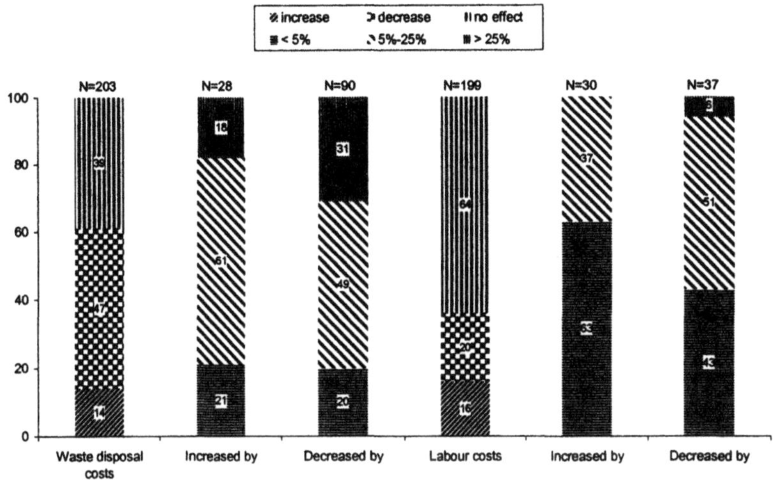

Again, we see that cost decreases dominate over cost increases, especially for waste disposal costs with 14% of the companies experiencing an increase in waste disposal costs and 47% experiencing a decrease as a result of the introduction of the innovation. For labour, 16% of the companies experienced an increase in costs, 20% experienced a decrease. The magnitude of the cost increase tends to be below that of the cost decrease for labour, so we also see here a synergy between cost reduction and environmental protection. The cost reduction for labour is somewhat surprising given the earlier finding that eco-innovation tends to create employment and increases qualification demand.

The cost effects are analysed statistically by the reasons for introducing the most environmentally beneficial innovation. The results are shown in Table 10, which gives the mean change in the number of employees and the mean percentage change in sales, energy costs, material costs, waste disposal costs, and labour costs. Firms that state that they introduced the innovation to comply with environmental regulations (compliance) have the lowest mean increase in the number of employees, a below average increase in sales, and a below average decrease in energy, material, and waste disposal costs. The only factor for which compliance firms do better is a slightly larger decrease in labour costs.

Table 10: Mean change in employment and costs for each reason for introducing the most beneficial environmental innovation

	Sales	Energy Costs	Material Costs	Waste Costs	Labour Costs
Compliance	2.4%	-3.9%	-1.0%	-4.0%	-1.1%
Secure existing markets	4.6%	-2.7%	-1.5%	-3.9%	-.03%
Increase market share	6.4%	-3.3%	-2.6%	-4.5%	-0.6%
Reduce costs	2.8%	-5.0%	-4.8%	-7.8%	-1.6%
Improve firm's image	2.7%	-3.9%	-2.2%	-5.7%	-1.1%
Respond to competitor	6.8%	-4.1%	-1.4%	-4.3%	0.7%
Achieve accreditation	3.1%	-4.8%	-1.8%	-5.2%	-1.1%
All firms	2.5%	-4.0%	-2.0%	-5.1%	-0.8%

10 The influence of organisational innovation

Organisational innovation should play an increasing role in innovation, due to the importance of investment in advanced manufacturing technologies based on information technology. This section takes a closer look at firms that report an organisational component to their most environmentally beneficial innovation. What is the effect of organisational innovation on sales, employment and costs? Does organisational innovation help companies to find cost reducing measures? In addition to differences by sector, several other firm characteristics were evaluated for organisational and non-organisational innovators: the cost of the innovation, the percentage of all innovation expenditures due to the innovation, whether or not the firm had received a subsidy to develop the innovation, if the innovation had been previously implemented by another firm, and who developed it. The results are given in Table 11.

Table 11: Characteristics of firms with and without an organisational component in their most environmentally beneficial innovation

	Organisational component		
	No	Yes	P
Cost in €			
< 50,000	43.0%	49.1%	
50,000-0.5 mln	37.1%	30.9%	
0.5-5 mln	10.6%	9.1%	.61
> 5 mln	2.0%	5.5%	
Don't know	7.3%	5.5%	
Percent of total innovation costs			
< 5%	30.5%	38.2%	
5% - 25%	22.5%	14.5%	
25% - 50%	15.9%	12.7%	.65
> 50%	13.9%	16.4%	
Don't know	17.2%	18.2%	
Received a subsidy			
No	74.8%	85.5%	
Yes	25.2%	14.5%	.07
Previously implemented by another firm?			
No	45.0%	45.5%	
Yes	55.0%	54.5%	.96
Who developed it?			
Your firm	29.1%	34.5%	
Other firm or organisation	34.4%	23.6%	
Your firm in cooperation with another	27.8%	41.8%	.06
Other/don't know	8.6%	0.0%	

There is no difference between organisational innovators and other innovators in the cost of the innovation, its share of total innovation costs, and whether or not it had been previously implemented. However, organisational innovations were less likely to receive a subsidy and more likely to have been developed in cooperation with another firm or institute.

The close linkages between organisational and other innovations makes it impossible to look at the employment effects of organisational innovation alone, with only one firm in this group. Any employment effect from the use of an 'organisational' innovation could also be due to another component, such as the process, product, service, or recycling category. Therefore, the organisational component is only an 'extra', with an unknown contribution.

There is no statistically significant difference by organisational innovation in the percentage of firms that state that the innovation increased, decreased, or had no effect on employment. 15.4% of organisational innovators reported an increase in employment vs. 14.4% of non-organisational innovators, 5.8% reported a decrease vs. 1.4%, and 78.8% reported 'no change' vs. 84.2% of non-organisational innovators. However, organisational innovators experienced a slightly greater mean change in employment of +2.4 employees vs. +0.5 for non-organisational innovators ($p = 0.06$). As noted above, however, it is difficult to know if this difference is due to the organisational component of the innovation or another category, particularly since the averages are dominated by relatively large employment gains in only a few firms.

What about the cost consequences? Does organisational innovation help companies to find and exploit cost reducing measures for energy, material, waste disposal, and labour costs? For this question we used information about the cost consequences from adopting the innovation. Respondents were asked to indicate whether they experienced a change in either the costs of energy, materials, labour, or waste disposal. If an increase or decrease was reported for these costs, the respondent was asked if the effect was less than 5%, 5%-25%, or greater than 25%. For analysis, mean values were assigned to each of the first two categories (positive 2.5% and 15% respectively for an increase and negative 2.5% and 15% respectively for a decrease) while the latter value was set to plus or minus 25%. Mean changes were then calculated for organisational and non-organisational innovators. The results are given in Table 12.

Table 12: Effect of organisational innovation on costs

Cost	Organisat. component[1]	Up	Down	No change	Total	Mean
Energy	No (147)	12.9%	40.1%	46.9%	100%	-3.5%
	Yes (54)	5.6%	44.4%	50.0%	100%	-5.3%
Materials	No (145)	22.1%	33.1%	44.8%	100%	-1.7%
	Yes (54)	24.1%	35.2%	40.7%	100%	-2.8%
Waste disposal	No (148)	11.5%	45.9%	42.6%	100%	-4.8%
	Yes (55)	21.8%	50.9%	27.3%	100%	-6.1%
Labour	No (146)	13.7%	18.5%	67.8%	100%	-0.9%
	Yes (53)	20.8%	24.5%	54.7%	100%	-0.8%

[1] Numbers in parentheses give the number of firms after excluding firms that replied 'don't know'.

In all analyses, there is a small negative decline in costs, ranging from -0.8% for labour costs to -6.1% for waste disposal costs, but none of the differences in the means between firms that include and do not include an organisational innovation are statistically significant. There is also no significant difference in the distribution of firms reporting that costs 'increased', 'decreased', or 'remained unchanged' for three of the four cost factors. The exception is for waste disposal costs, where more firms with an organisational component report an *increase* in costs. This is not supported by the mean differences in costs, however, where there is no difference.

11 Process innovation and end-of-pipe replacement

To analyse the employment effects of process innovations vs. end-of-pipe technologies, the survey inquired into the effects of a shift from end-of-pipe to process innovation. Process innovators that did not introduce a product or service innovation were asked if the process innovation replaced a previous process or an end-of-pipe pollution control system. The question excludes product and process innovators since these could be introducing a new type of process linked to the innovation. Of the 137 process innovators, 81 did not introduce a product or process innovation and therefore responded to the question. The process innovation replaced an existing process in 35 firms (43%) and an end-of-pipe system in 24 firms (30%). Of these, 15 stated that the process innovation both replaced part of an existing process and an end-of-pipe system.

Table 13 provides the employment effects for process innovators that did not report product or service innovations. None of these firms reported a decrease in employment. There are no significant differences in employment changes among the three groups, although firms that used the process innovation to replace an end-of-pipe innovation had the smallest impact on employment.

Table 13: Change in employment among process innovators

	N	Employment increased	No change	Mean change in employees
Process innovators but no replacement	36	11.1%	88.9%	0.17
Replaced process only	19	10.5%	89.5%	4.4
Replaced end-of-pipe[1]	23	4.3%	95.7%	0.04

[1] 15 of these firms also report replacing a process innovation.

The results indeed show that the shift from end-of-pipe to process change mitigates the positive employment from process innovation but does not result in a net reduction of employment. In the majority of cases, however, it did not make a difference for employment.

12 Basis of competition

The respondents are asked to identify which of five factors are an 'important basis of competition between your establishment and your competitors'. The question refers to *all* of the firm's activities. The five factors include price, quality, environmentally friendly features, innovative products or services, and corporate image. The importance of 'environmentally friendly features' is of greatest interest here. A second question asks the respondent to identify which of these five competitive factors are the *most* important to the firm.

On average, 3.8 of the five factors are cited by each firm. Figure 13 gives the distribution of competitive factors by sector, summed across all citations. Environmentally friendly features is in the last place for both manufacturing and service firms, accounting for 13.1% of the total citations. There are only minor differences by sector. The only significant difference is for price, which is more frequently cited by manufacturing firms. In absolute terms, price is the most frequently cited competitive factor, noted by 87% of the firms, while environmentally friendly features are cited the least frequently, by 49.8% of the firms.

Figure 13: Sector distribution of the factors of competition

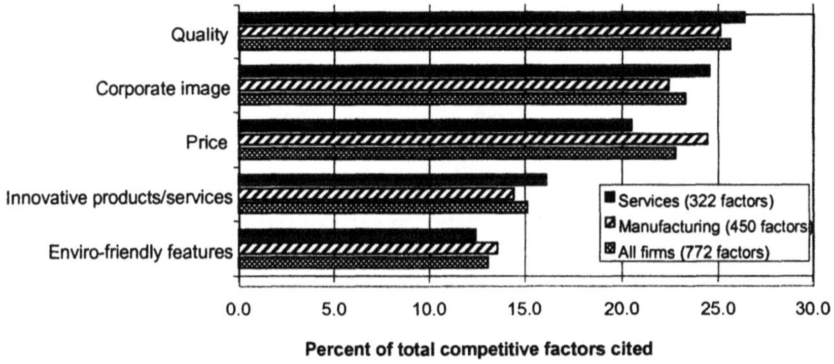

The importance of 'environmentally friendly features' falls precipitously when the respondents are asked to choose the most important basis of competition, as shown in Table 14, where only 1.5% of the respondents mark this factor as the most important one. The almost non-existent role of environmentally friendly features makes it impossible to evaluate possible correlations between product innovation and this factor of competition. Instead, such a comparison must be made using the responses to the previous question on the factors that form an 'important' basis of competition.

Table 14: Percent of firms citing at least one factor as the most important basis of competition

	Manufacturing	Services	All firms
Price	52.2%	32.6%	43.8%
Quality	33.9%	39.5%	36.3%
Environmental features	1.7%	1.2%	1.5%
Innovative products/services	3.5%	8.1%	5.5%
Corporate image	8.7%	18.6%	12.9%
Total	100%	100%	100%

Table 15 evaluates the relationship between each of the six types of environmentally beneficial innovations that were introduced by the firm in the previous three years and 'environmentally friendly features' and 'innovative products and services' as a basis of competition. No results are given for the other three competitive factors since they are too commonly cited to result in many differences by innovation type. The only significant difference by whether or not a firm introduced a specific type of innovation with environmental benefits is for organisational innovations. 37.2% of firms that did not introduce an organisational innovation cited 'environmentally friendly features' as a basis of competition, compared to 59.0% of firms that did introduce an organisational innovation. Of interest, whether or not the firm introduced a product with environmental benefits has no effect on the use of either environmental features or innovative products/services as a basis of competition. This is very surprising, as we would have expected a positive relationship.

Table 15: Environmental innovation and competition on the basis of 'environmentally friendly features' and 'innovative products/services'

		Basis of competition:			
Environmental Innovation introduced?		Environmentally friendly features		Innovative products/ services	
		Cited	p[1]	Cited	P[1]
Organi-sational	No	37.2%		58.1%	
	Yes	59.0%	.00	57.3%	1.0
Pollution control	No	44.3%		53.6%	
	Yes	54.7%	.16	61.3%	.32
Recycling	No	44.0%		62.6%	
	Yes	54.5%	.16	53.6%	.20
Products	No	48.9%		54.3%	
	Yes	50.5%	.89	60.4%	.40
Processes	No	42.1%		56.1%	
	Yes	52.7%	.21	58.2%	.88
Distribution	No	51.8%		54.4%	
	Yes	47.2%	.36	61.8%	.32

[1] Significance level of difference.

13 The effects of environmental regulation on competitiveness

Two questions in the questionnaire ask the respondent to estimate the importance of environmental regulations on the 'types of processes used by your establishment" and on the 'types of products or services that your establishment can produce and sell'. The question refers to *all* of the firm's activities and not just to its most environmentally beneficial innovation. Four options are given: 'very important', 'moderately important', 'no effect', and 'don't know'. For these analyses, a response of 'don't know' is assumed to be equal to 'no effect', since otherwise the manager should be aware of regulatory effects. A third question is asked, on a yes or no basis, if the 'firm changed its products or processes specifically to comply with environmental regulations'.

Table 16 gives the distribution of responses by sector. Environmental regulations have a greater impact on processes than on products, with only 13% of manufacturing firms stating that regulations had no effect on processes while 28% said that regulations had no effect on products.

There is no difference by sector in the effect of regulation on the types of products sold, but manufacturing firms are significantly more likely to find that regulations have an effect on processes, with 44% of manufacturing firms stating that regulations had a 'very important' effect on their processes, compared to only 30% of service firms. Slightly more manufacturing than service firms stated that they had changed a product or process in response to regulations, but the difference is not significant. There was no difference by firm size (large vs. small) in the distribution of responses on the regulation (results not shown).

Table 16: Importance of environmental regulations for products and processes among 117 manufacturing and 91 service firms

	Sector	Very impor-tant	Moderat. impor-tant	No effect	P[1]
Types of products sold	Manuf.	35%	37%	28%	
	Services	36%	28%	36%	.30
Types of processes used	Manuf.	44%	44%	13%	
	Services	30%	37%	33%	.002
Change in response to regulation			Yes	No	
	Manuf.		59%	41%	
	Services		52%	48%	.33

[1]Difference in the distribution of responses between manufacturing and service firms.

Overall, over half of the firms have changed a product or process specifically to comply with an environmental regulation, although the effect of regulation is more marked on processes than on products.

Table 17: Regulation, sales and employment

	Mean change in sales			Mean change in employment		
	Manuf firms	Service firms	All firms	Manuf firms	Service firms	All firms
*Importance of **product** regulations on the types of products and services produced and sold*						
Very important	7.8%	14.8%	10.9%	3.3%	23.7%	12.2%
Moderately important	9.6%	14.6%	11.5%	1.9%	12.1%	5.5%
Not important	8.7%	15.1%	12.0%	3.4%	45.0%	24.2%
P	Ns[1]	ns	ns	ns	ns	Ns
*Importance of **process** regulations on the types of processes in use*						
Very important	10.0%	17.1%	12.4%	5.7%	30.2%	14.2%
Moderately important	8.6%	13.1%	10.5%	0.2%	19.0%	7.6%
Not important	3.7%	15.2%	11.2%	1.3%	36.7%	25.5%
	Ns	ns	ns	ns	ns	Ns
*Firm changed processes or products **specifically to comply** with environmental regulations*						
No	11.7%	18.1%	14.7%	1.5%	44.7%	21.9%
Yes	6.7%	12.1%	8.9%	3.6%	12.5%	7.2%
P	.10	ns	.04	ns	.07	.07

[1] ns = not significant.

A common concern on the part of industry groups, policy makers and economists is that regulations can interfere with competitive performance outcomes such as sales, employment and profits. A preliminary investigation of this issue can be made by looking at the respondent's perception of the severity of regulation and the change in total sales and employment in the previous three years[2]. The results are given in Table 17.

The importance of regulation has little effect on total sales and employment. For example, the average change in sales for manufacturing firms that report that

[2] Please note that it is not very meaningful to look at the effect of regulation on sales of products influenced by the firm's most important environmental innovation, since there is no information linking regulation to this particular environmental innovation.

regulations have had a 'very important' effect on the 'types of products and services produced and sold' is 7.8%, which is not significantly different from the change in sales of 8.7% for manufacturing firms that stated that regulations were 'not important'. Whether or not the firms changed either a process or product 'specifically to comply with environmental regulations' has a greater effect. Generally, sales and employment grew faster among firms that replied 'no'. This result suggests that regulation does have a negative effect on output indicators, but it is curious that there is no effect for the severity of the regulations.

14 Regression results for sales and employment

The effect of environmental activities on the firms' sales and employment in the Netherlands is investigated through a series of multivariate regressions. Two methods are used. First, an OLS regression is used for the percentage change in each firm's total sales and total employment. Second, a logistic regression is used to explore the factors that influence whether or not the firm's most environmentally beneficial innovation increased total unit sales and the number of long-term employees.

The first set of regressions concerns the percentage change in total employment and sales. General firm characteristics that could influence the change in total sales and employment are firm size, measured as the log of the number of employees (LEMPLOY), the firm's sector of activity (MANUF), and the percentage of all employees with a university education (UNIEMP).

Environmental variables include those that are limited to the firm's most environmentally beneficial innovation and those that concern general activities. The former group could have only a weak influence on the change in total employment and sales, although the effect should increase with either the total amount invested in the innovation (MCOST) or the share of all innovation expenditures spent on this innovation (SHARE). In addition, firms that have introduced an eco process innovation (PROCX) could experience a decline in employment, while firms that introduced an eco product or service innovation (PRODSERV) could experience an increase in employment. Any possible changes due to process or product innovation could have little to do with the environmental component and only reflect general effects from process vs. product innovation. Another environmental indicator is whether or not the firm introduced the innovation in order to comply with environmental regulations (COMPLY). The concern is that this type of innovation could reduce sales or employment.

Several variables reflect the firm's general environmental activities. These include 'environmentally friendly features' as an important basis of competition

(ECOCOMP) and if the firm has changed a process or product specifically to comply with environmental regulations (REGCHAN).

Table 18 provides the regression results for both the OLS and logistic regressions. The OLS results have very weak R^2 values, so the results need to be interpreted cautiously. The logistic models provide stronger results, but the model only estimates the likelihood that the firm reports an increase in employment (vs. no change or a decline) and an increase in sales (vs. no change).

Manufacturing firms in the OLS regressions have lower percentage changes in employment and sales, which confirms the results, shown in Table 3, that service firms are experiencing faster growth rates. Firm size (LEMPLOY) has no effect. The share in total investment of the firm's most environmentally beneficial innovation (SHARE) has no effect on employment and could not be entered into the model for sales, due to poor fit. Of interest is that a product or service innovation (PRODSERV) increases sales, as predicted, while a process component to the most environmentally beneficial innovation (PROCX) both reduces employment and sales. Firms that have changed a product or process in response to regulations (REGCHAN) have lower sales growth but this variable has no effect on employment. These results confirm the finding for sales in Table 17 but refute the findings for employment.

Neither the firm's sector (MANUF) nor its size (LEMPLOY) have any effect in the logistic regressions. The log of the total amount spent on the firm's eco-innovation (LMOST) provides a better fit than SHARE and increases the probability of both an increase in employment and sales. As with the OLS results, a product or service innovation (PRODSERV) increases employment and sales, but process innovation has no effect.

Please note that compliance with environmental regulation does not influence either sales or employment, which contradicts the results shown in Table 10. Introducing the innovation in order to comply with environmental regulations has no effect either. Overall, the most important factors that determine whether or not the eco-innovation increases employment and sales are the amount spent on it (confirming the results in Figure 7) and whether or not the innovation has a product or service component.

Several other variables were included in the regressions, such as whether or not the innovation was developed within the firm, but these variables had to be excluded because they introduced collinearity problems into the regressions.

Table 18: Regression results for employment and sales

	OLS for percent change in total employment and sales		Logistic for an increase from the eco-innovation	
	Employment	Sales	Employment	Sales
Constant	17.0 (17.8)	8.9 (7.4)	-6.0 (1.9)	-4.1 (1.5)
MANUF	-11.6 (6.7)	-4.8 (2.9)	-.62 (.57)	-.31 (.49)
LEMPLOY	.80 (7.7)	4.2 (3.4)	-.43 (.66)	-.08 (.58)
SHARE	-1.03 (2.8)			
LMCOST			.68 (.28)	.42 (.18)
PRODSERV	7.4 (6.5)	4.8 (2.9)	1.7 (.59)	1.7 (.51)
PROCX	-12.2 (6.6)	-6.6 (2.9)	.81 (.62)	.16 (.50)
ECOCOMP	4.0 (6.3)	.7 (2.8)		
COMPLY			-.70 (.52)	-.56 (.46)
REGCHAN	-6.6 (6.3)	-5.7 (2.8)		
UNIEMP	.44 (.18)	.1 (.1)	.01 (.01)	.01 (.01)
No. Firms	164	164	156	156
Model signif.	.007	.003	.000	.000
R^2	.13	.12	.38	.29

15 Conclusions

The average Dutch eco-innovative firm introduced around 4 different types of environmental innovations, with process innovation being cited by 71% of the respondents, followed at a distance by organisational innovation (57%) then recycling (56%) or pollution control (52%). Process innovation is also viewed the most important environmental innovation (cited by 24.2% of all companies and 29.7% of manufacturing firms). Second often cited is pollution control (19.8%). Product innovation comes fourth, after recycling. This shows that in protecting the environment, there has been a shift away from pollution control (end-of-pipe technologies), towards process integrated measures and to product change and recycling. End-of-pipe solutions still constitute an important category. Most of the innovation types are more commonly cited by manufacturing firms, with the exception of organisational and service innovations which are more frequently mentioned by service firms.

Eco-innovations (at least the most significant ones) do not fit the categories commonly used for describing environmental innovation: pollution control, process change, recycling, product innovation, service innovation, organisational innovation and new distribution system. Only 24 firms introduced just one of the six possible environmental innovations.

The focus of much of the questionnaire is on the firm's most environmentally beneficial innovation, which can be one of eight different innovation types, or a combination thereof. The innovation could have been introduced specifically to solve environmental problems or the environmental benefits could have been unintended – a positive side-effect. The fact that all but 21% of the respondents cited multiple innovation types illuminates the complexity of innovation, but at the same time this result makes it very difficult to link a particular type of innovation to a particular outcome of interest. Further work in this area is required to disentangle the effects of different innovation components.

There are significant differences by the firm's sector in the amount spent on the most beneficial eco-innovation. Manufacturing firms spent, on average, five times as much on this innovation than service firms. But, they were also more likely to receive a subsidy or grant. Investment costs were highest for process innovation in manufacturing which had a mean value of 4 mln €, twice the investment cost of the average innovation. For 61% of the companies, the eco-innovation amounted to more than 5% of the innovation expenditures, and for 36% more than 25%. It shows that firms spend a comparatively large fraction of their total innovation investment on innovations with an environmental benefit, although for some of these firms the environmental gain can be unintended.

22% of the firms received a subsidy or grant for their innovation, with the highest subsidy rate among large manufacturing firms (27%) and the lowest rate for small service firms (20%).

With a few exceptions, there is very little difference in the motivation behind each innovation component. Innovations with a product or service component are more likely to be introduced to secure existing markets or increase market share, while process innovations are more likely to be introduced to comply with regulations. The former two reasons, not surprisingly, are positively correlated with an increase in sales. Introducing an environmental innovation to comply with environmental regulations has the least beneficial effects to the firm in terms of employment, sales and a reduction in input costs. On average, 3.4 reasons were cited by 205 respondents as being important for adopting/developing their most important environmental innovation. Only 24 firms cited just one reason. This shows the multiplicity of motivations for eco-innovation.

The most notable result in terms of employment is that for a large majority of firms, eco-innovation has no discernible effect on employment, with 82% of the respondents citing 'no change'. The highest frequency of an increase in employment is for innovations with a product or service component. Positive employment effects are also more prevalent among firms that developed the innovation themselves, but this can be partly explained by the fact that firms tend to develop product innovations in-house and outsource their process and pollution control innovations to other firms – probably capital equipment suppliers. The cost of the innovation is also positively correlated with positive employment benefits, an unexpected result.

Almost 60% of firms that found their eco-innovation to increase employment reported a need for new skills, compared to only 25% of firms that stated that their eco-innovation had no effect on employment. Most firms met their demand for new skills through training new employees (87%), followed by outsourcing (56%).

Overall, the effects of the eco-innovation on energy costs, material costs, waste disposal costs and labour costs turn out to be favourable; shares of companies experiencing a reduction in costs are higher than those experiencing an increase in costs, with decreases in costs exceeding the increases in costs. The eco-innovation helped to decrease energy costs for 41% of the companies and reduced material costs for 34%. 47% of the companies experienced a reduction in waste disposal costs due to the innovation. This suggests that on the whole there is a synergy between environmental protection and cost reduction. This corresponds with findings from other studies (for example, Dieleman and de Hoo, 1991, Clayton et al., 1999 and studies mentioned in Howes et al., 1997).

The survey asked about five general competitive strategies. Almost all firms cited quality, corporate image and price, while slightly less than 50% cited 'environmentally friendly features'. When asked to identify their most important basis of competition, only 1.5% cited environmental features. Clearly, environmental features, although a background part of many firm's innovative activities, is rarely a major competitive factor. The only innovation component that is significantly correlated with environmental features is organisational innovation.

Environmental regulation raises concerns about a loss of competitiveness in terms of sales or employment. Firms that had introduced a product or process specifically in response to regulation had a significantly lower mean rate of sales and employment growth, compared to firms that did not. The severity of regulation for products and processes had no effect, however, while the effect on employment does not hold in the regression analyses.

The multivariate regression results confirm many of the descriptive results that have been discussed so far. There are also a few changes. The differences between manufacturing and service firms no longer hold in the regressions, indicating that other measurable factors, such as a focus on product vs. process innovation, or the cost of the innovation, lie behind the sectoral differences.

Several in-depth analyses of these data are required in the future. First, discriminant analysis needs to be used to differentiate between the different components of the firm's most environmentally beneficial innovation. Second, the effect of the background rate of employment and sales growth on differences due to eco-innovation needs to be examined and clarified.

References

Clayton, A., G. Spinardi and R. Williams (1999), *Policies for Cleaner Technology. A New Agenda for Government and Industry*, European Commission, London.

Dieleman, H. and S. de Hoo (1991), Towards a Tailor-made Process of Pollution Prevention and Cleaner Production: Results and Implications of the PRISMA project, in: Fischer, K. and J. Schot (Eds.), *Environmental Strategies for Industry: International Perspectives on Research Needs and Policy Implications*, Washington, D.C., 245-276.

Garrod, B. and P. Chadwick (1996), Environmental Management and Business Strategy: Towards a New Strategic Paradigm, *Futures* Vol. 28, No. 1: 37-50.

Howes, R., J. Skea and B. Whelan (1997), *Clean and Competitive? Motivating Environmental Performance in Industry*, London.

Kemp, R. and A. Arundel (1998), *Survey Indicators for Environmental Innovation,* IDEA paper, No. 8, STEP 1998.

Williams, H. E., J. Medhurst and K. Drew (1991), Corporate Strategies for a Sustainable Future, in: Fischer, K. and J. Schot (Eds.), *Environmental Strategies for Industry: International Perspectives on Research Needs and Policy Implications*, Washington, D.C., 117-146.

The UK survey

Steven Glynn

1 Description of the sample

The UK sample consists of 400 firms selected to reflect the UK economy as a whole – with the proviso that only firms with more than 50 employees were considered. The sample was selected using Dun & Bradstreet Marketplace Tool Ver 2. – October 1999. In realising the proposed sample, there were problems in obtaining sufficient interviews with, in particular, small service firms. Hence the use of a weighted share, to bring the achieved sample back into line with the originally planned sample, and make comparisons with other surveys more feasible.

All of the firms interviewed undertook some form of environmental innovation within the last three years – firms that did not were screened out. We thus have a sample of eco-innovators rather than a sample of UK industry in general. Table 1 outlines the broad structure of this sample in terms of sector and size.

2 Environmental innovation

Figure 1 shows all the innovations that were introduced by the sample of firms, in the last three years, that were regarded by the respondents as having an environmental benefit. From the range of innovations that they had undertaken, firms were then asked to select which was the most environmentally beneficial. In classifying their innovations, respondents to the interview were not restricted to one category.

Table 1: Description of the sample

	Number of firms	Share	Weighted share
Small	254	63.50	72.75
Large	146	36.50	27.25
Industry/manufacturing	211	52.75	43.00
Service	189	47.25	57.00
Manufacturing	162	40.50	33.04
Electricity, gas and water	11	2.75	2.11
Construction	38	9.50	7.85
Wholesale/retail trade	74	18.50	26.39
Hotels and restaurants	5	1.25	1.69
Transport, storage and communication	56	14.00	14.23
Financial intermediation	10	2.50	3.22
Real estate, renting and business activity	44	11.00	11.46

While we can see that a broad range of environmentally beneficial innovations have been introduced in the last three years, it is clear that, for the UK, **recycling** innovation is seen as both the most common and most environmentally beneficial type of innovation. Of the other innovation types, perhaps the most striking result is that while organisational innovation is the second most mentioned innovation, it is far less likely to be portrayed as the most environmentally beneficial. This result deserves further investigation as to the characteristics of such innovations.

Figure 1: Environmentally beneficial innovation

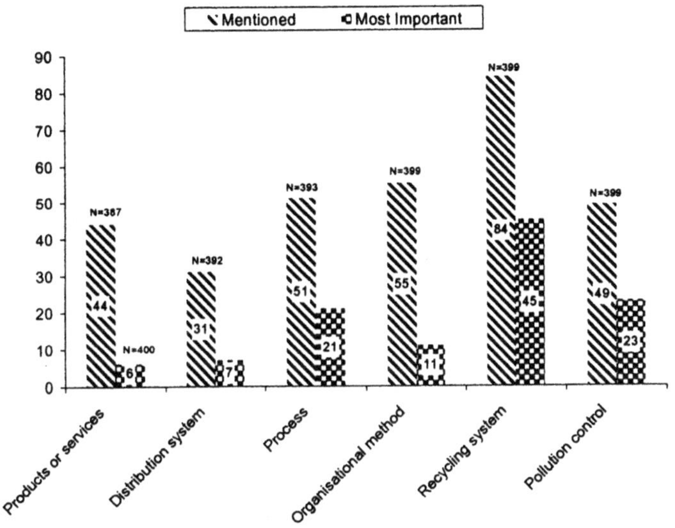

Figure 2 details the most important reasons for introducing an innovation. Three things in particular stand out. Firstly, "improving the firm's image" is the most important reason given for introducing the environmental innovation concerned, followed by "reducing costs" (although this is less important for product innovations). Meanwhile, both "responding to a competitor's innovation" and "achieving accreditation" are seen as unimportant reasons for the innovation in question – although see below for an exception concerning accreditation. A further feature that bears mentioning is the similarity in the distribution of responses concerning securing and increasing market share.

A number of points also arise regarding the specific innovation types. In terms of the reasons given, the innovation type does seem to be important, with considerable variation between the different types. Only for the "improving firm's image" category is there a fairly even distribution of the innovation types. One of the most marked differences is in the "achieve an accreditation" category where organisational innovation stands out from the others – usually accreditation is a minor objective. One possible reason for this is that organisational innovation arises due to firms seeking ISO 14001 or similar accreditation.

Figure 2: Important reasons for introducing the innovation

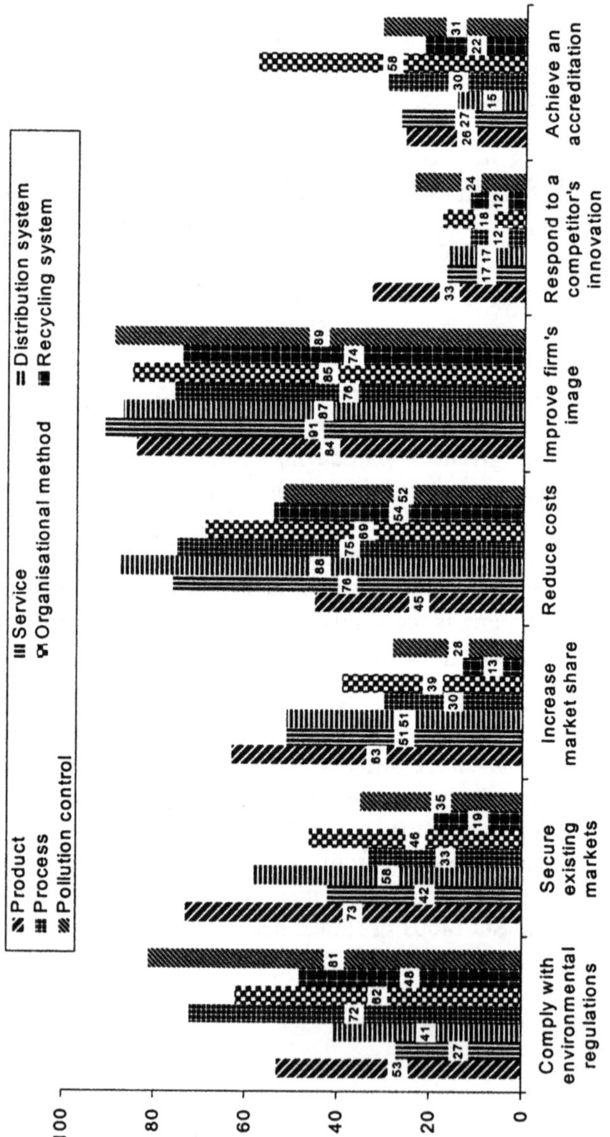

Moving on to the impacts of eco-innovation, the most obvious point to draw from Figure 3 is that the environmental innovations that are considered here did not result in much employment change – with only 5% of respondents saying that employment increased and 2% saying that it had decreased as a result of the innovation.

So employment impacts are infrequent, and when they do occur, they more often involve increases than decreases. Pollution control innovation is the only category that was more likely to lead to a decrease in employment than an increase. None of the service, organisational or recycling innovations led to an employment decrease. We consider issues of skills a little later.

Figure 3: Effect of innovation on employment

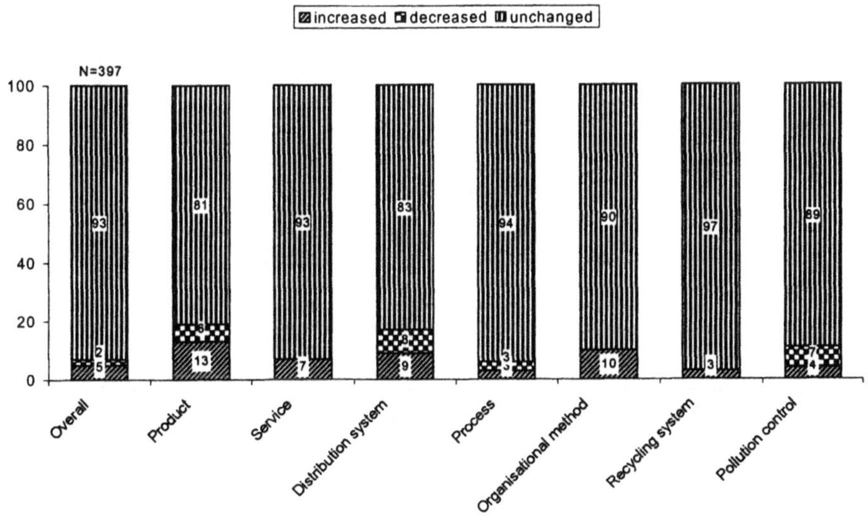

Figure 4 examines the origins of the specified innovations. The majority of firms had themselves some involvement in developing the innovation, either alone or with another firm(s). A minority of roughly one third of firms (34%) simply acquired an innovation from external sources. It is interesting to note that where employment changed (in either direction) the innovation was more likely to have been developed by the firm itself. This is, perhaps surprisingly, slightly more the case for employment decreases, although where employment increased it is much less likely that the firm had no involvement in the innovation. However, thinking back to Figure 3, we must be careful in interpreting these results as, when we refer to cases where employment increased or decreased we are discussing only a small number of cases (20 for increase, 9 for a decrease).

The commentary here refers simply to what is apparent from visual inspection of the data, and makes no claim as to statistical significance (i.e. how likely it would be that such tendencies would be apparent in the larger population from which our sample is drawn).

Figure 4: Who developed the innovation

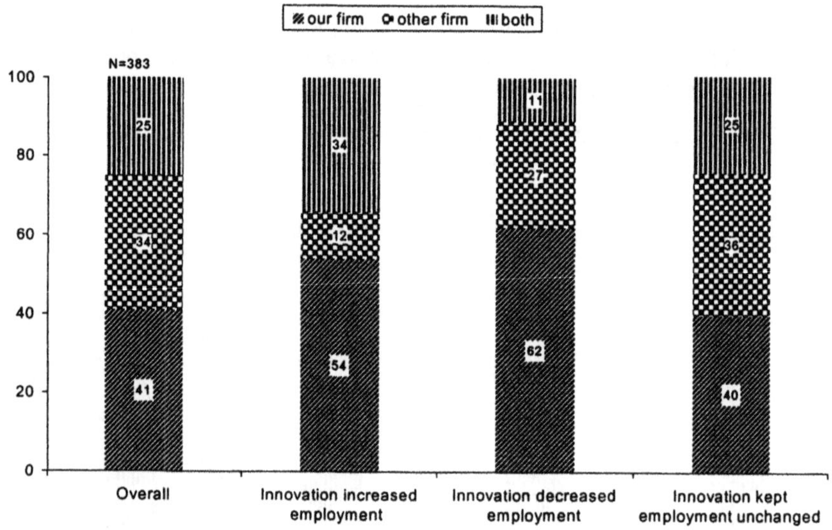

Figure 5 gives an overview of the investment costs that were involved in undertaking the most environmentally beneficial innovation. Overwhelmingly, firms invested rather little – less than 50,000 € (82%) – on the innovation. Perhaps the high proportion of recycling innovations (Figure 1) could provide an explanation for this – these innovations may often require little more than the purchase of some storage containers and the devotion of a small amount of staff time. It is more surprising that **none** of the cases when employment increased involved an investment of more than 5 mln €; in contrast those innovations that decreased employment were more likely to be high cost ones and rarely the very cheap ones. (The previous proviso regarding the few cases where employment increased or decreased applies here also.)

Figure 5: Investment costs of the innovation

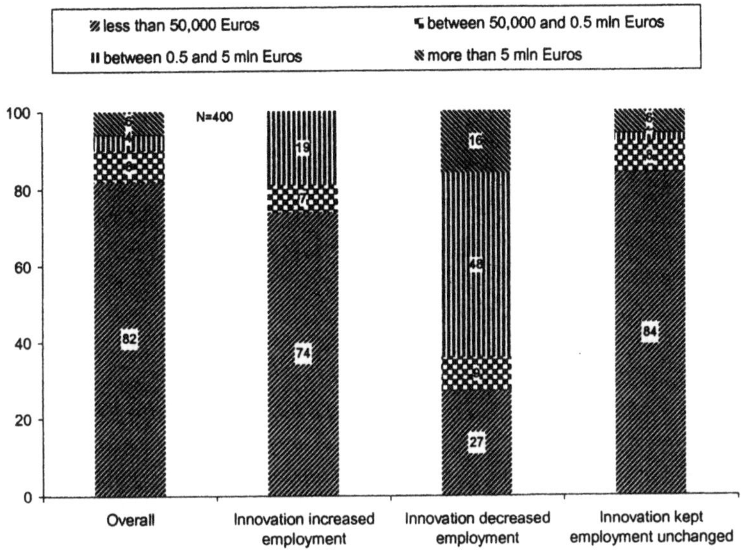

Figure 6: Proportion of innovation expenditures

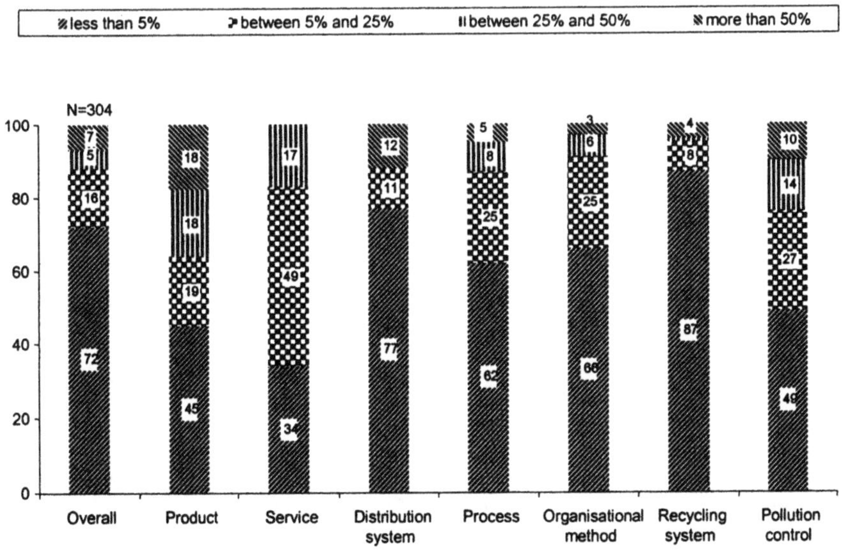

In looking at the proportion of innovation expenditure in Figure 6 the innovation types appear to break down into three groups. For a very high number of cases of *distribution* and *recycling* innovations expenditure was low – less than 5% of total innovation expenditure. For *process* and *organisational* innovations there are more cases where a higher proportion of the innovation budget was spent on the innovations in question, although the greatest share of cases still involve less than 5% of the budget. However, for *product*, *service*, and *pollution control* innovation, the proportion of innovation expenditure is generally over 5% – these are clearly the more financially demanding innovations.

Looking in more detail, two points stand out. Firstly, there are no cases for *service* innovation where the proportion of innovation expenditure was over 50%, despite the fact that these innovations are also the least likely to involve very low levels of expenditure. Secondly, despite a generally low proportion of innovation expenditure being spent on distribution systems, there are a relatively high number of innovations of this type that took over 50% of the overall innovation expenditure. This bimodal structure suggests that while many distribution system innovations may be small-scale, some are very significant – one possible example could be where firms replace fleets of vehicles, install expensive logistics systems, etc.

Figure 7: Was subsidy/grant received (by innovation type)

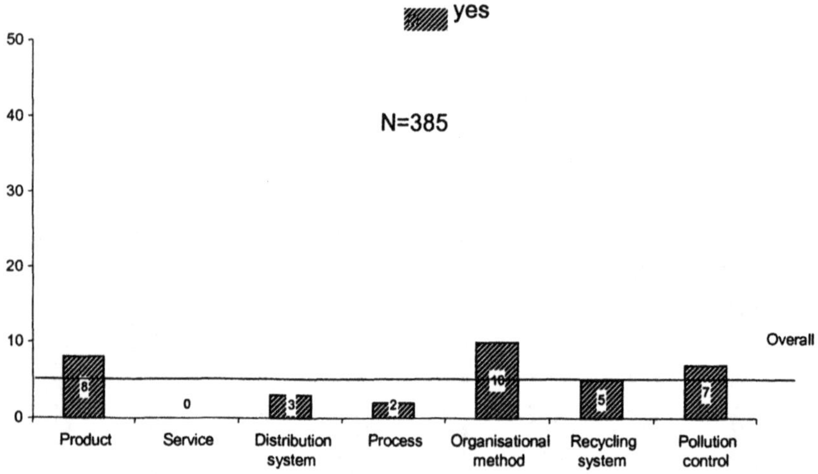

What about sources of financing: Are there incentives from public sources? The main conclusion that we can draw from Figure 7 is that very few UK firms received a subsidy to undertake what was their most environmentally beneficial innovation. The relatively high level of subsidy for organisational innovation bears further investigation – we suggest that it could be as a result of more local/regional initiatives, perhaps again related to the attainment of ISO 14001 (e.g. subsidised training courses in local colleges).

Just under 20% of firms felt that the innovation introduced required new skills Figure 8 allows us to address two questions. The first concerns whether the need for new skills is affected by the innovation type; the second, the relationship between the need for new skills and the tendency for employment to increase, decrease or remain stable. Regarding the first question, recycling innovation is particularly unlikely to require new skills. Given the high number of cases falling into this category, this has clearly brought down the average, as is clear from the other innovation types – though the greatest requirement for new skills is still only 40% for organisational innovation. Regarding the second question, it is clear that where the innovation has resulted in employment change there is a far higher requirement for new skills. This is particularly marked for the cases where employment decreased – perhaps where employees are lost the remaining employees need new skills to cover for that loss, or perhaps few high skill employees are substituting for many low skill ones.

Figure 8: Need for new skills by innovation type and employment effect

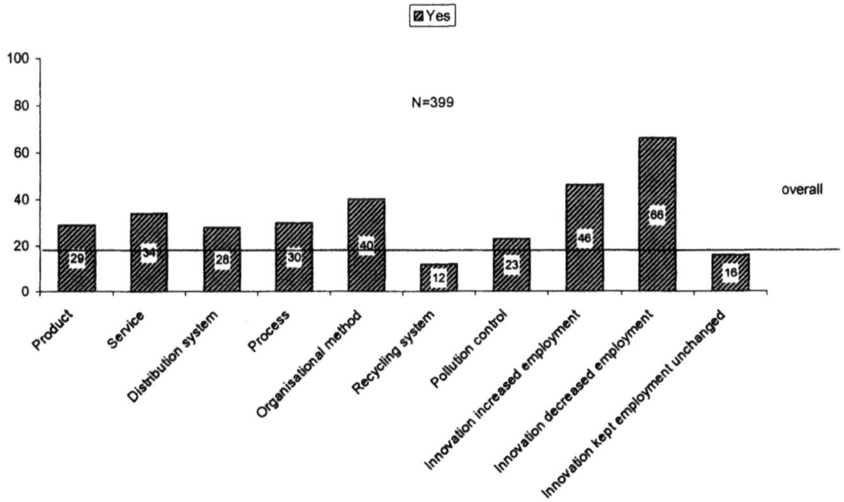

Where new skills were required, Figure 9 shows that the most common method of meeting the need was, unsurprisingly, to train existing employees. Considering only the firms that said the innovation did require new skills, the figure for training existing employees is very substantial – around 90%. Of the other methods of meeting the skills, temporary hiring is least likely with firms perhaps preferring to outsource work rather than take on temporary staff to meet temporary needs.

Figure 9: How were new skills met

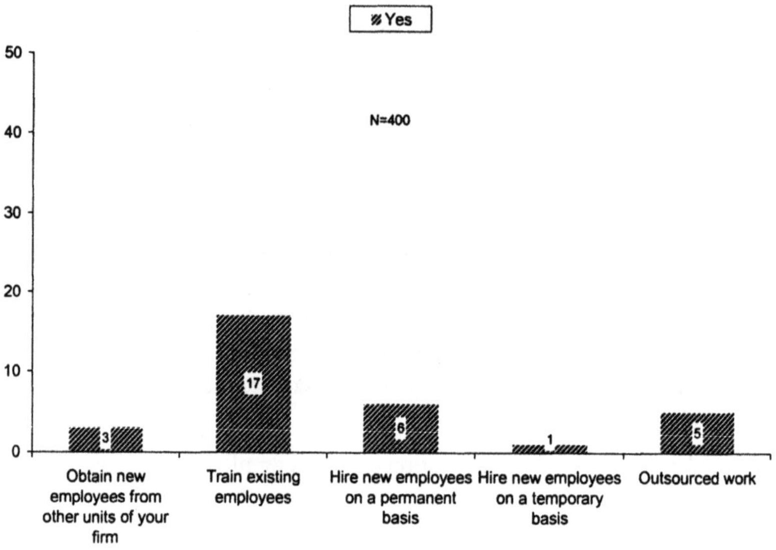

Figure 10 looks at which types of employees received training, when firms did acknowledge the need for retraining. The pattern for service firms and manufacturing firms is similar, with skilled workers the most likely to receive training, followed by unskilled workers, and lastly those workers with a university degree (probably managers and professional staff). Where the innovation resulted in employment change, either increase or decrease, all firms undertaking training retrained skilled workers. However, it is interesting to note that where employment decreased, other employee types were far less likely than average to receive training. (Again, caveats as to the small numbers involved.) The fact that more skilled and unskilled employees are trained is perhaps encouraging – it suggests that for the environmental innovations discussed here, skills are not being polarised.

Figure 10: Who received training by industry and employment effect

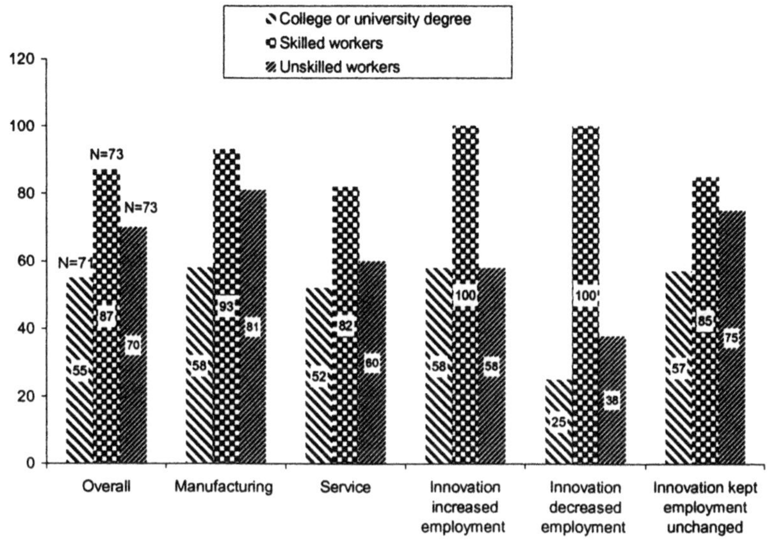

Figures 11a, 11b and 11c show the effects of the innovation in question on sales, prices and energy, material, waste disposal and labour costs.

Figure 11a: Effects of innovations on sales and prices

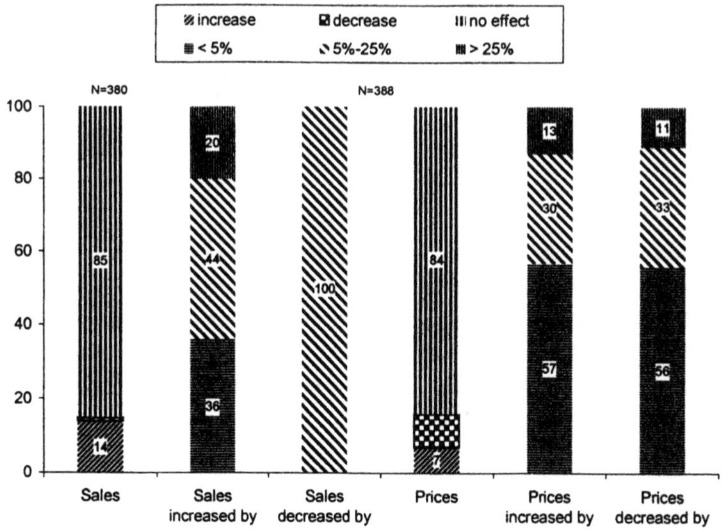

Regarding sales and prices, the results resemble those for employment. In most cases the innovation had no effect. Where there was an effect on sales it tended to involve an increase, which could be quite large. There is a very small number of cases reporting decreases (3 firms), so it is probably unwise to read much into the 100% of sales decreasing by 5-25%. For prices, the innovation was slightly more likely to result in a decrease than an increase, although the changes are generally quite small.

For both energy costs and material costs (Figure 11b), there were around twice as many cases where the innovation decreased costs than those registering increases, although it was again most common to report no effect. In addition, decreases in costs tended to be more significant that increases in costs.

Figure 11b: Effects of innovations on energy and material costs

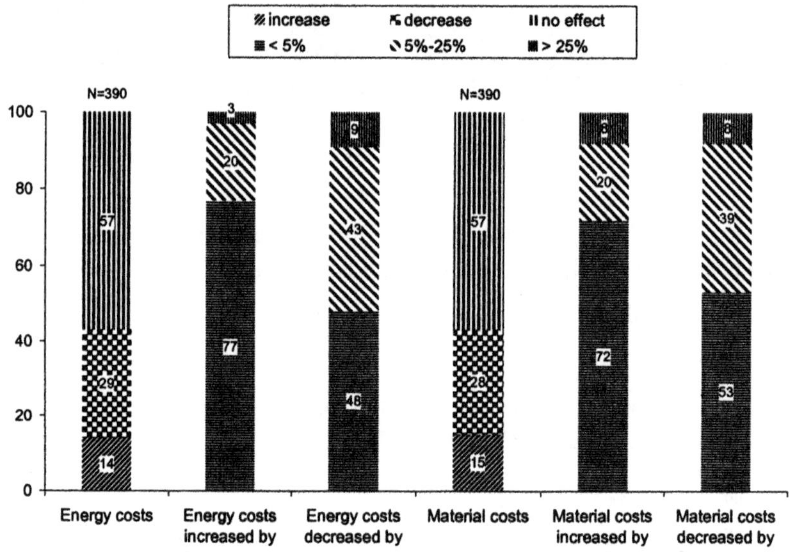

Figure 11c shows that for labour costs, too, no effect was most likely. However, waste disposal costs run counter to this pattern. A high proportion saw increases or decreases, with the latter being most significant, both in terms of numbers of cases and in terms of the amount by which the costs decreased. The fact that waste disposal costs are the most affected could again be related to the high numbers of innovations that were seen as recycling, although there are likely to be impacts from other innovations too.

Figure 11c: Effects of innovations on waste disposal and labour costs

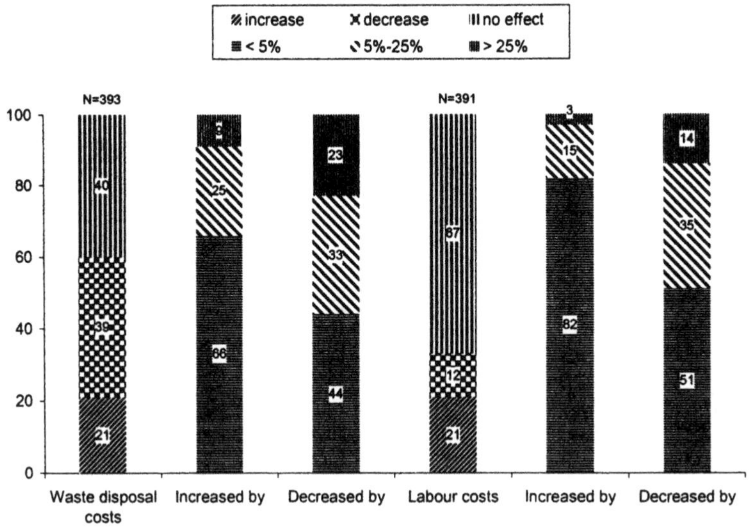

3 General questions concerning the firm

In this section, the focus moves away from the most environmentally beneficial innovation to focus on more general questions about the firm's environmental stance.

Figure 12 shows that environmental regulations are more important for processes than they are for products. In other words, the regulations hit the way in which firms produce their goods and services (and thus create "externalities") to a greater extent than they lead to changes in the market outputs themselves. The final column concerns whether firms have had to make changes in their products or process over the last three years, specifically to comply with environmental regulations. It shows that in the last three years just over 50% of the firms sampled have been driven to make changes due to environmental regulations. The sample of eco-innovators, then, is almost equally split between those where change has been driven by particular regulation, and others.

Figure 12: Importance of environmental regulations for processes and products and changes to comply with regulations

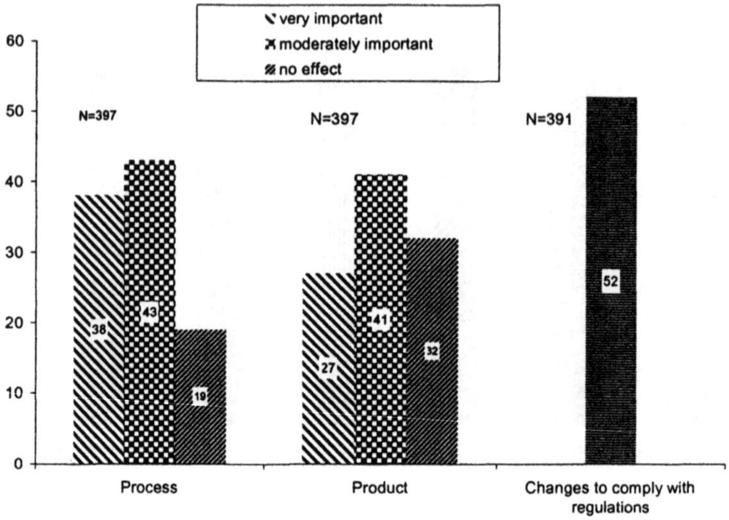

Figure 13 displays firms' views as to important factors in competition, and, among these, which factor was considered the most important.

It is clear that as well as being most frequently mentioned, *price* and *quality* are seen as by far the most important factors in competition. While both *innovative products or services* and *corporate image* are frequently mentioned, few see them as the most important. Strikingly, given the focus of the study, *environmentally friendly features* are mentioned relatively infrequently – and are seen by only 2% of firms as the most important factor.

In terms of changes in overall employment, rather than employment change due to the specific innovation, there are no surprises revealed in Figure 14. Overall, the environmentally innovative firms considered in this study increased employment. Small firms have seen more of an employment increase than big firms, and employment in service firms has increased much more than employment in manufacturing firms. Looking at firms where employment changed due to the specific innovation that concerned us previously, it is clear that firms that increased employment due to the innovation were more likely to grow overall, while those that decreased employment tended to be decreasing employment overall.

Figure 13: Important competition factors

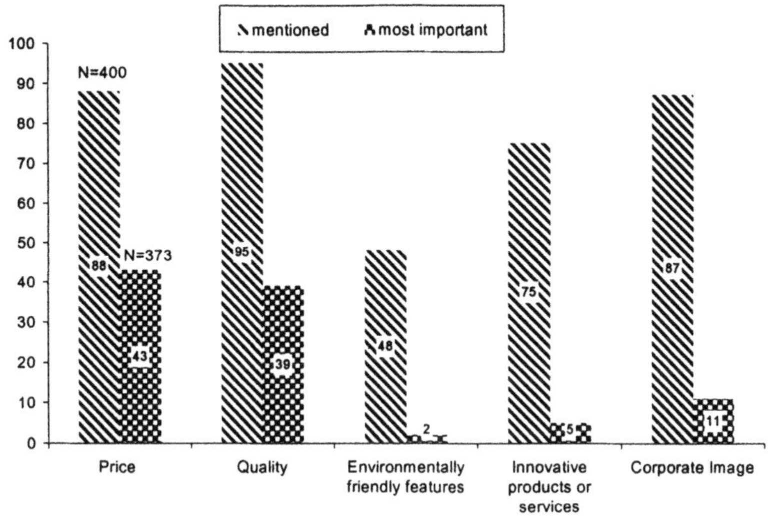

Figure 14: Change in overall employment

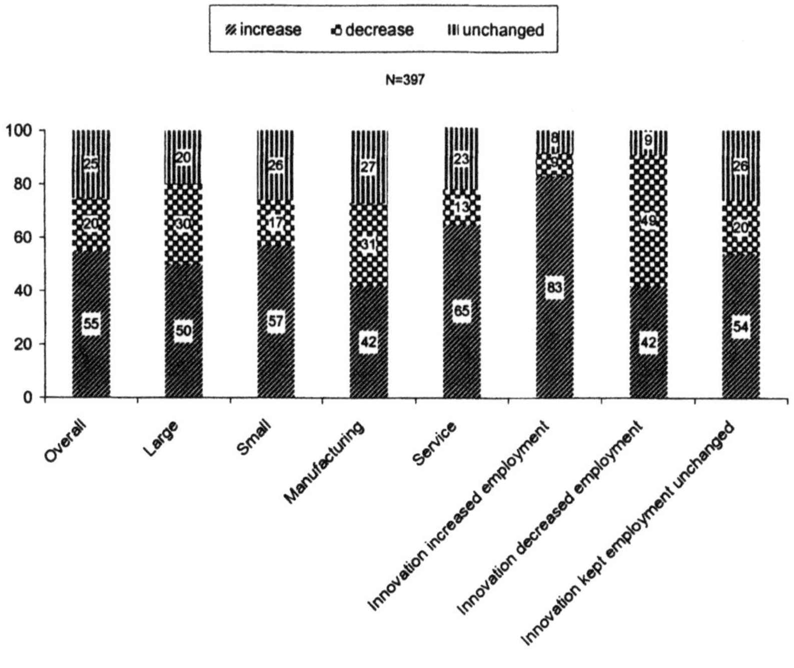

Figure 15 displays the proportion of companies that offer products or services that are specifically marketed as environmentally friendly, within groups reporting different employment effects. The firms that signalled an employment change due to the specific innovation were more likely to have specific environmentally friendly products and services than average – some 60% of those reporting positive change, and 50% of those with job losses. However, given the earlier proviso about the number of these firms, this result is really in need of further investigation. For now, perhaps the main point to take from this figure, and reflecting what was said about Figure 13, is the relatively low number of firms that do offer environmentally friendly products.

Figure 15: Does your company offer products or services that are specifically marketed as environmentally friendly

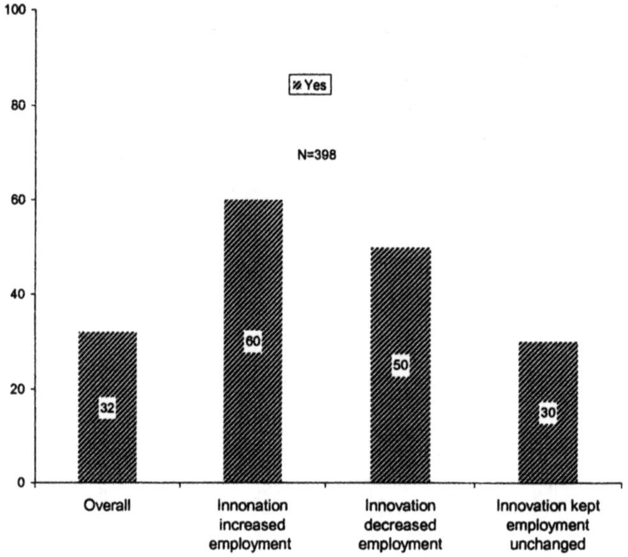

The Italian survey

Matteo Bartolomeo and Rita Canu

1 Description of the sample

The universe of Italian establishments has been derived from data collected by Istituto Nazionale di Statistica (ISTAT, the National Statistical Office) through the Censimento Intermedio dell'Industria e dei Servizi (Intermediary Census on Industry and Services) that has been recently published. The definition of the sample has been implemented with reference to the structure of industry and service sectors as they were at December 31st 1996.

The IMPRESS survey for Italy considered 8 cells, created according to 3 dimensions (see Table 1): geographical location (North vs. Centre-South); size of the business (from 50 to 199 employees and 200 and more employees) and kind of activity (industry vs. service).

According to the Census, for the 8 cells considered in the IMPRESS survey, there are more than 23,000 establishments situated in Italy. Total employment is around 3.5 mln. More than 14,000 of them are active, at 31 December, 1996, in manufacturing (2.2 mln employees), while the number of the establishments active in the service sectors is around 9,000 with 1.3 mln employees.

Due to the limited size of most of Italian firms, the exclusion of firms with less than 50 employees has dramatically reduced the reference universe and therefore the sample: this represents just 0.6% of establishments active in the sectors considered. Employees working in establishments that are part of the universe considered by the IMPRESS survey are 26% of total employees in sectors taken into account (69% in electricity, water and gas distribution; but only 7% in construction and wholesale/retail trade).

The Italian sample has been also specified in terms of geographical location of the establishments considered. These have been divided in two different groups: North of Italy (Piemonte, Valle d'Aosta, Liguria, Lombardia, Trentino-Alto Adige, Veneto, Friuli-Venezia Giulia, Emilia Romagna) and Centre-South of Italy (Toscana, Umbria, Marche, Lazio, Abruzzo, Molise, Campania, Puglia, Basilicata, Calabria, Sicilia, Sardegna). The distincion has been considered relevant for the well-known difference in the distribution of activities between the two geographical areas. Around 2/3 of innovative companies are located in just four regions of the North (Piemonte, Lombardia, Veneto, Emilia Romagna). Considering investments for innovation, 71% of them are also concentrated in the North. The result complies with similar findings of a survey conducted in 1985.

The geographical distinction between North and Centre-South, also underlines that 66% of establishments are located in the North, with quite a difference between industry (71%) and service (59%).

Also in terms of employment, the distribution among North and Centre-South follows a similar structure (56% north and 44% Centre-South)

In order to carry out the 400 interviews, a stratified sample has been extracted from the universe with a sampling fraction of around 0.017. The extraction has been implemented randomly from the Registro delle Imprese (Register of firms) updated in 2000. The definition of the activity code is normally carried out by the personnel of Chambers of Commerce (where the Registro delle Imprese is maintained) while the number of employees is declared by the firms themselves on a regular basis. These discrepancies, together with the fact that the data set, with data on the establishments, used for conducting the survey is updated on 31 December, 1998, has induced researchers to clean the final data base with interviews. 16 records have been deleted, thus reducing the total number of interviews (400) to 384. This difference then created some minor errors in the proportionality of the sample that required the attribution of weights as indicated in the right column of Table 1.

The total response rate has been close to 15%, relatively high in Italy for such a survey.

Table 1: Description of the sample

	Number of firms	Share	Weighted share
Small	321	83.59	84.59
Large	63	16.41	15.41
Industry/manufacturing	247	64.32	61.41
Service	137	35.68	38.59
Manufacturing	222	57.81	54.67
Electricity, gas and water	9	2.34	2.65
Construction	16	4.17	4.09
Wholesale/retail trade	31	8.07	8.69
Hotels and restaurants	10	2.60	2.32
Transport, storage and communication	50	13.02	13.01
Financial intermediation	10	2.60	5.59
Real estate, renting and business activity	36	9.38	8.99
North Italy	258	67.19	66.16
Centre-South Italy	126	32.81	33.84

2 Environmental innovation

Figure 1 shows survey results for innovations introduced and for the most environmentally beneficial innovation. In the last three years the majority of establishments interviewed have introduced pollution control and process innovations with benefits for the environment (59 and 58%). For many of them this innovation is also the most important environmental innovation adopted (45 and 44%).

Interesting to notice is that product and service innovations have also been introduced quite extensively, but just in a few cases these innovations were the most important one (15% of cases) also because the demand for environmentally sound products and services has also a limited role.

These results can be explained by the orientation of Italian legislation on environmental performance of production processes (to be addressed via pollution control measures, new process and, partly, recycling systems). On the opposite, the concept of extended producer responsibility – which would entail product, service and alternative means for distribution/logistics – is still at an infancy stage.

Figure 2 illustrates the reasons for introducing the innovation. With regard to the most important environmental innovation introduced by the establishment, the motivations refer primarily to assuring compliance with regulations and improving the firm's image.

Figure 1: Environmentally beneficial innovation

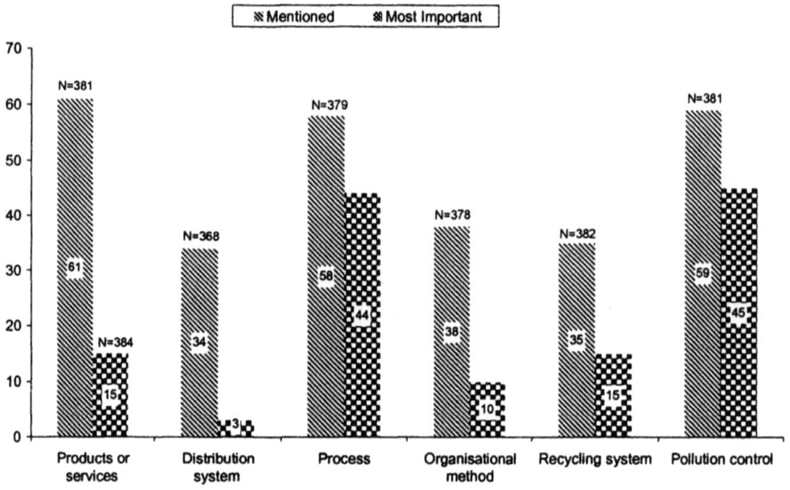

Figure 2: Important reasons for introducing the innovation

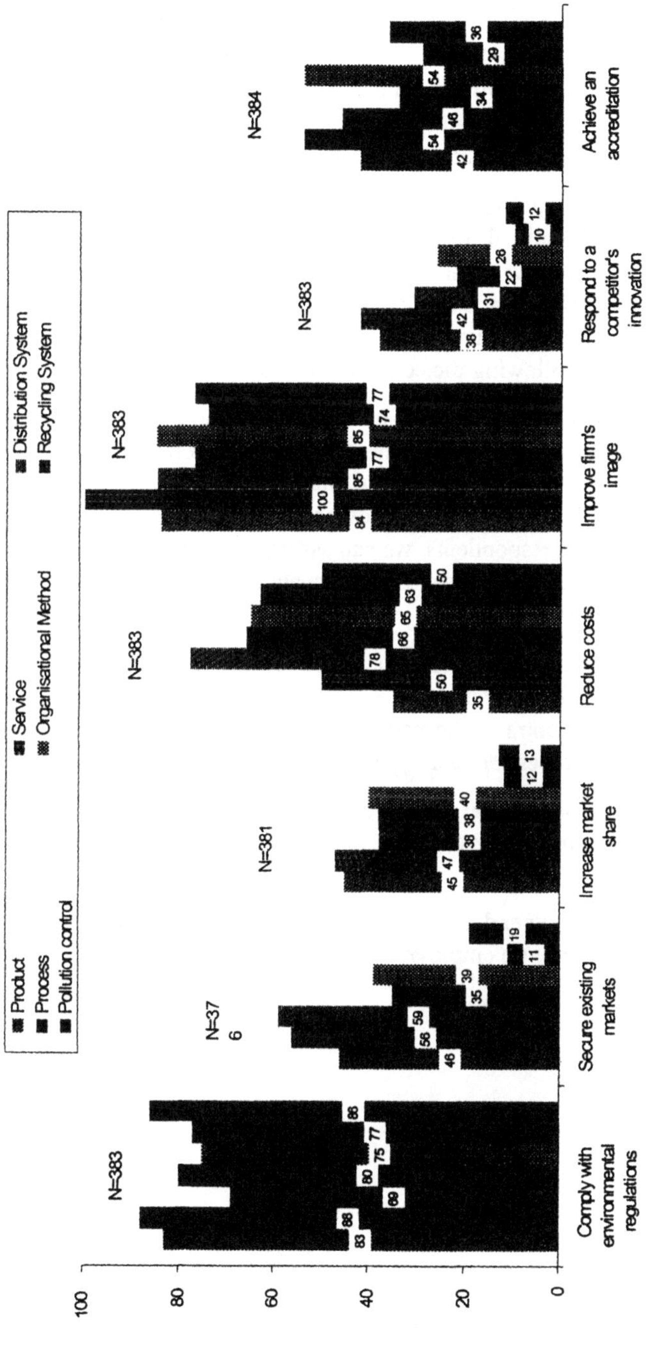

Other motivations related to competition related issues – such as secure existing markets, increase market share or respond to a competitor's innovation – are by far less important.

The target of an environmental certification, such as ISO 14001 and the EU EMAS scheme, is apparently less important but may have some overlaps with the objective of improvement of image and respond to competitors' innovation.

Regarding the typology of the innovation contributing to the achievement of the objectives, virtually all innovation types increase firm image and in address environmental regulation. Pollution control and service innovation offer the contribution to meet environmental standards. When we move to more competitive issues, product, service, organisational method and distribution system are the most effective ones in allowing the company to reach the goal.

Figure 3 considers the effect of innovation on employment. It is important to stress that respondents, the majority dealing with environment-related tasks, have provided information on the basis of their personal knowledge and feeling for the issue considered. Since the employment effect is rather far from traditional competencies of the respondents, we can assume that some mistakes could occur in the information collected and that employment effects are not only due to the adoption of the innovation considered.

The overall employment effect is rather positive for the innovation considered in the survey with on average 11% of cases where there is an increase in employment. On the contrary, the negative employment effect is normally registered in around 1% of cases for each type of innovation considered.

Results demonstrate that the effect is more positive for "soft" innovations, such as service, distribution system and organisational method, which are more labour-intensive.

Quite surprisingly end-of-pipe (pollution control) innovations have a very limited positive influence on employment.

Figure 3: Effect of innovation on employment

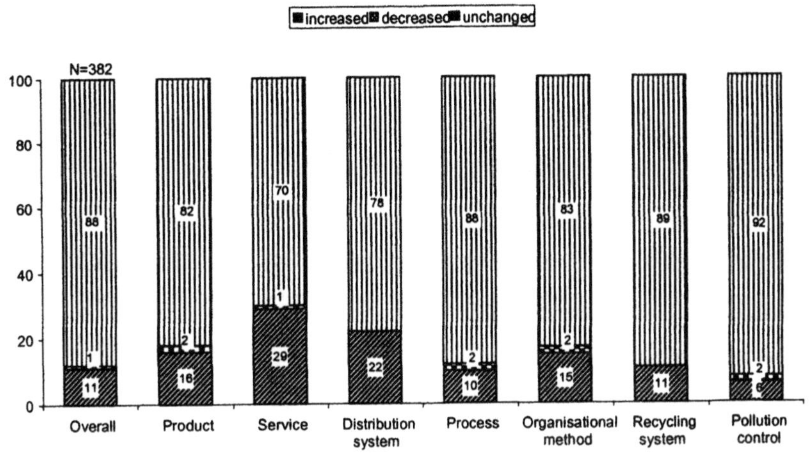

Figure 4 shows the relationship between the developer of the innovation and the employment effect generated by the innovation. As a general remark, the most important environmental innovation considered has been developed by other firms for the majority of cases analysed (64% of cases analysed), while only in 30% of cases the innovation has been developed internally. This can be explained by two different arguments:

- On the one side the firms interviewed are in the large majority small and medium sized enterprises (around 85% of the total sample), with a relatively weak ability to develop innovations by themselves.
- On the other side, the most important environmental innovations introduced, which are the only ones considered for this and the following questions, are more related to technology than to "soft" measures; the ability to innovate is then even weaker.

Having said that, it is difficult to explain why there is more frequently a decrease of employment when the innovation is developed internally or jointly. One explanation, which is more a speculation rather than a scientific evidence, could be that innovations developed internally have more frequently the objective of reducing costs and therefore labour costs. A more basic explanation could refer to the insufficient number of observations, especially as far as the increase of innovation is concerned.

Figure 4: Who developed the innovation

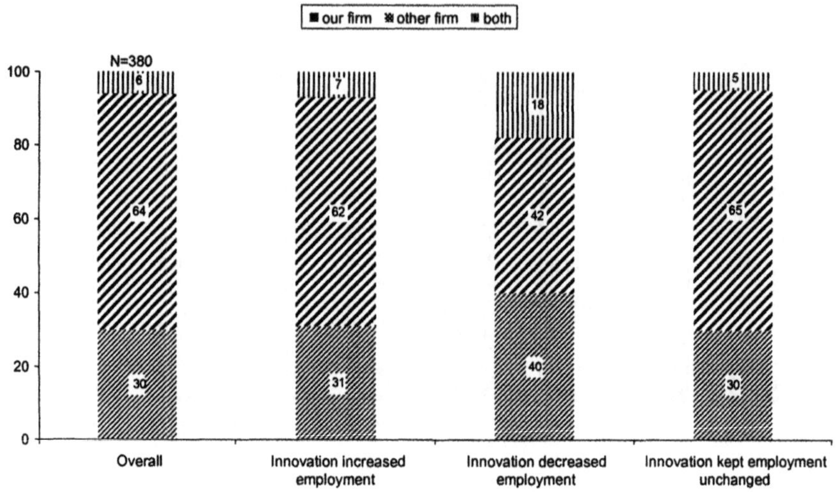

Figure 5 provides an interesting insight in the relationship between investment costs of the innovation and the employment effect. More precisely, when the cost is relatively limited, the employment effect is positive while, when the costs increase, the employment effect decreases.

The possible explanation for these results is that in the former case the innovation consists more often of new organisational methods, services development or distribution systems, the more labour-intensive innovations. In the latter case, the investment required for the innovation is relatively high, the innovation is more related to equipment, machines and plants that have a negligible or negative impact on innovation. These results are very consistent with those presented in Figure 3, showing a more positive relationship between employment increase and "soft" (and therefore less costly) measures.

Differently, when the investment costs of the innovation are limited, the employment effect is negligible or not easy to be assessed by the interviewed person.

Figure 5: Investment costs of the innovation

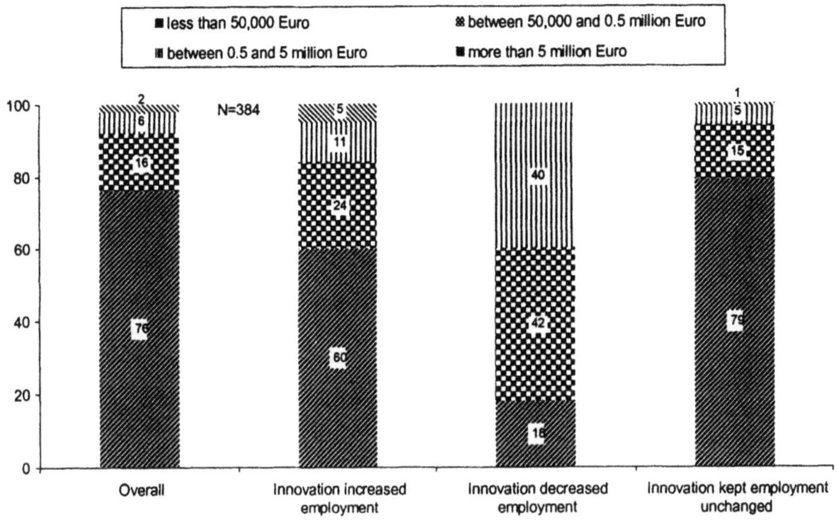

Figure 6 shows the proportion of expenditure for the innovation considered (the most important environmental one) on the total sample and with reference to the innovation type.

The vast majority of cases is characterised by an expenditure which is below 25% of total innovation costs borne by the establishment. Just in 9% of cases the innovation costs exceed 50% of total innovation expenditure.

Recycling systems absorb just a small portion of innovation costs, followed, surprisingly, by pollution control systems. An explanation for data on pollution control can be provided by considering that pollution control itself also includes monitoring systems and separate collection of wastes.

The most surprising situation refers to organisation method and to distribution systems: in both cases the overall cost for the adoption of the innovation is high (for organisation method, the costs exceed 50% of total innovation costs in 20% of cases). The limited number of observations for the distribution systems and organisational methods (as most important innovations) suggests that the sample is not large enough to draw conclusions.

Figure 6: Proportion of innovation expenditures

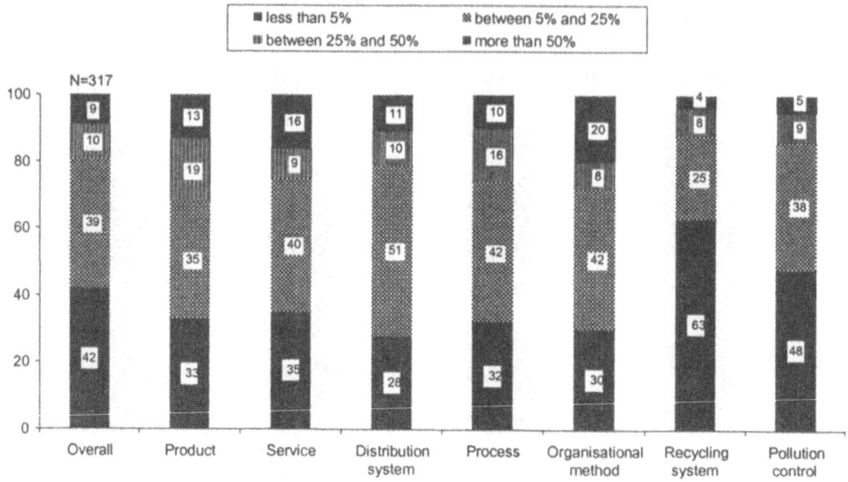

Figure 7 considers the relationship between receiving a subsidy and adopting a certain kind of innovation. In line with national and regional policies in recent years, innovations in terms of organisational methods have been quite extensive (17% of cases, above the average which is 11%). Similarly, national and regional policies have provided incentives for integrated innovations, namely process innovations in 15% of cases.

The figure on services are quite influenced by innovations regarding the substitution of engines and vehicles for transportation companies. In this case, the national policy for substitution of unleaded engines and obsolete vehicles has provided an easy platform for receiving subsidies at the moment of purchase.

Figure 8 considers the skill effect, by providing information on the need for new skills at the level of establishment. Results are matched with data on innovation type and on the direction of employment change (employment increased, decreased, no effect).

The IMPRESS survey demonstrates that different types of innovations require a very similar skill shift: for all kinds of innovation considered, a skill shift is required for a percentage of cases which is always included in a small range between 63 and 49%. The only exception is represented by organisational method, where new skills are needed in more than 80% of cases.

Figure 7: Was subsidy/grant received (by innovation type)

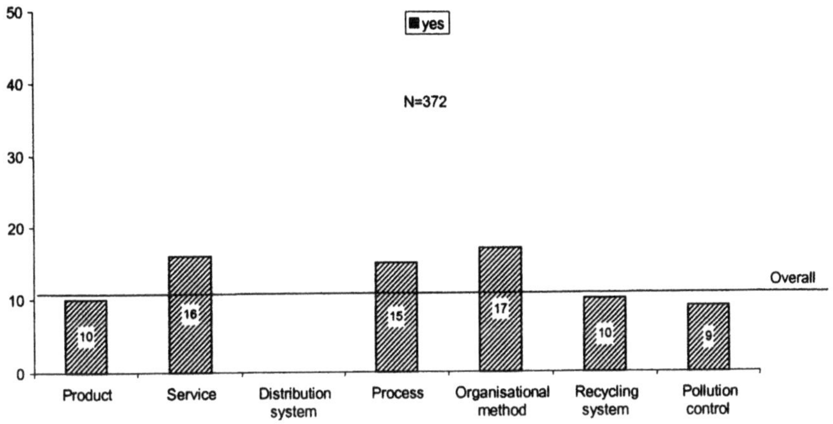

Figure 8: Need for new skills by innovation type and employment effect

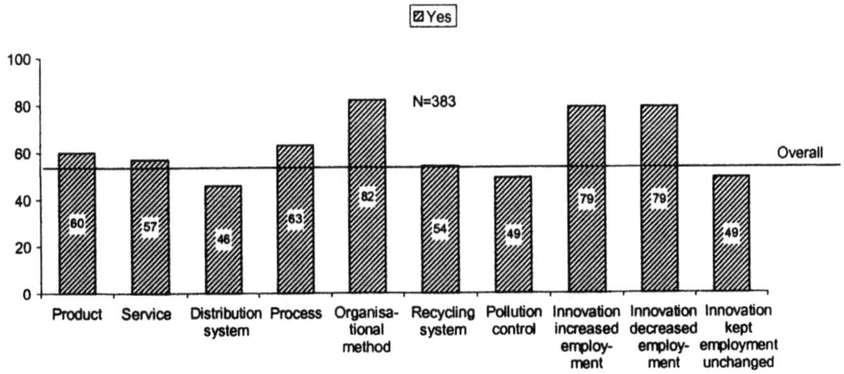

Figure 10 considers the issue of training and the addressees of training divided in three different categories of employees (unskilled, skilled, college).

Skilled workers are the most important target for training activities related to the adoption of the innovation (since tasks of employees with college and university degree and, on the opposite, unskilled employees remain relatively unaffected by the innovation), but the differences with other categories are not dramatically relevant.

Training is more important for service sector than for manufacturing. One can assume that this is due to the higher importance of organisational innovations, which require more skill adjustment in the service sector than in manufacturing.

It is interesting to notice that training activities occur more often when the innovation determines a decrease in employment. Two probable reasons are:

- because of the need for instructing employees to multi-tasking activities.
- due to the fact that, when employment increases, the new skills are acquired rather than created internally.

When employment decreases, both skilled and unskilled workers receive training courses in all situations. On the opposite, when the employment effect is positive, unskilled workers are the target of training activities only in 61% of observations.

Figure 9: Who received training by industry and employment effect

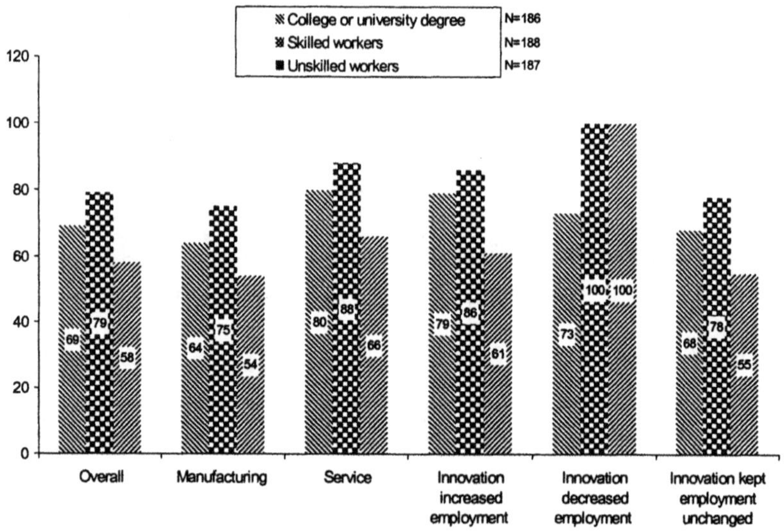

Figures 10a, 10b and 10c consider the effect of the innovation on sales, prices, energy costs, material costs, waste disposal costs, labour costs.

The innovation normally does not deeply affect prices and quantity of products/services sold in the market.

Energy costs are affected in nearly 50% of cases considered in the survey, 29% of which reported a quite tangible decrease of energy costs. A similar situation has been registered for material costs.

The innovation determines a quite considerable variation of waste disposal costs: around 50% of the establishments analysed have reported such a change, with a slight majority of cases where waste disposal costs increased. On the opposite, the extent of the variation is more considerable for the companies that have decreased their waste disposal costs.

Also in the case of labour cost, the innovation has a quite remarkable effect with roughly 30% of cases reporting a change even if the magnitude of change is by far below 5% of total labour cost.

Figure 10a: Effects of innovations on sales and prices

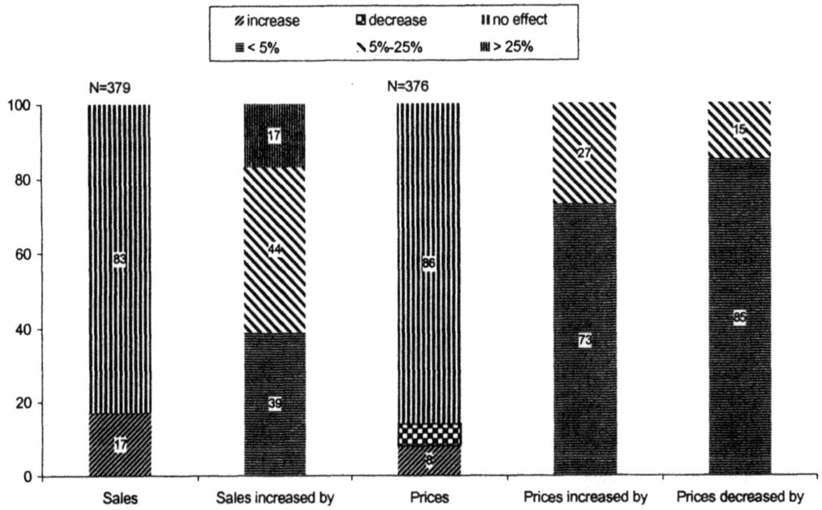

Figure 10b: Effects of innovations on waste disposal and labour costs

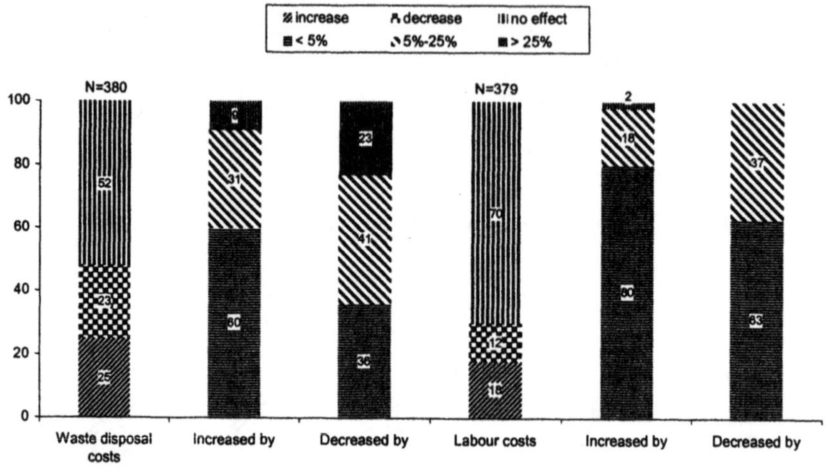

Figure 10c: Effects of innovations on energy and material costs

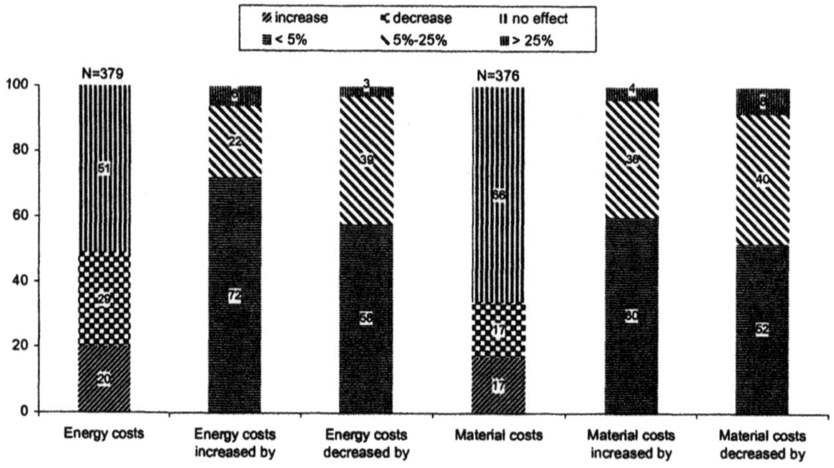

3 General questions

Figure 11 considers the way in which regulations affect innovations on product and processes made in the last three years.

The average result suggests that more than 50% of respondents consider environmental regulation as an important driver for innovation.

Clearly environmental regulation has been an important driver for process innovation and a slightly less important driver for product innovation. Differences are tangible but not particularly relevant.

Figure 11: Importance of environmental regulations for processes and products and changes to comply with regulations

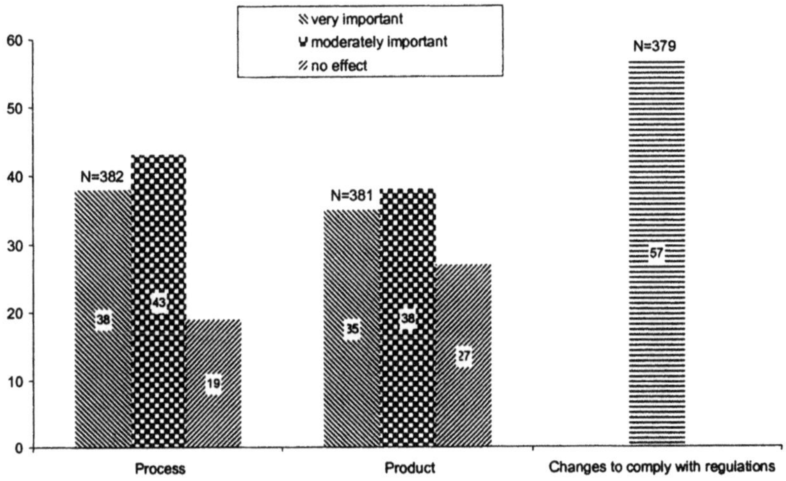

Figure 12 considers the general attitude of establishments towards competition by providing information on the most important factors for competition.

The results show the same ranking for the issues mentioned and those that are considered most important. Quality is the most important reason (54% of cases) and it is also most frequently mentioned (94% of cases).

Environmentally friendly features appear to be the least important reason among those mentioned by the interviewed. On the other side, it is self-evident that environmental features could contribute to increase quality of products, to build corporate image and to increase innovative activities. The importance of environment as competitive factor is therefore more implicit than explicit and the highest ranking for quality (as opposed to price) supports this argument.

Figure 12: Important competition factors

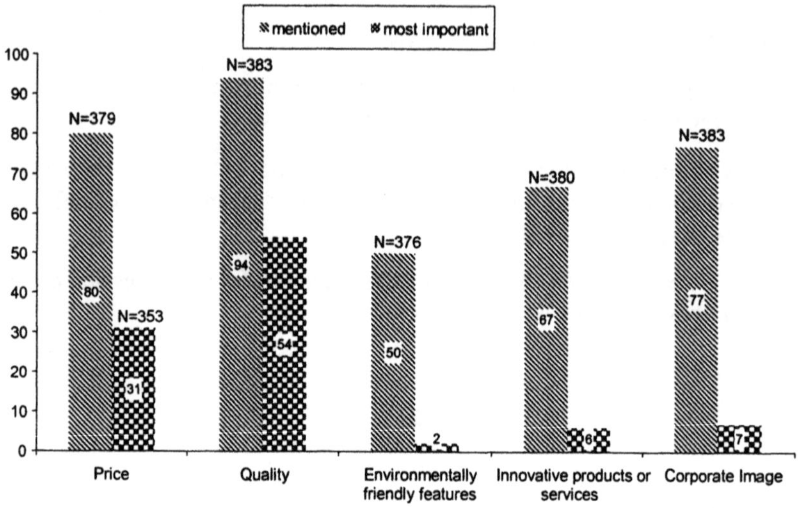

Figure 13 illustrates the changes in overall employment (without regard for the specific environmental innovation) for the establishments considered by dividing the sample into large and small, manufacturing and service.

The majority of establishments considered in the sample have recorded an increase in overall employment (52% increase, 11% decrease, 37% unchanged) thus suggesting that eco-innovators (which are the only ones considered in the sample) are successful at least in terms of growth of the business.

Small establishments demonstrate a better ability to increase employment; nevertheless the survey does not capture the extent of the change which could, in theory, offset the above mentioned result. Differences between manufacturing and service companies are quite negligible.

The right part of the figure provides an insight on the relationship between environmental innovation and related employment effects and, on the other hand, overall employment effects for the establishment considered. The results are very interesting: 92% of the establishments that reported an employment increase due to the eco-innovation also registered a general positive employment effect. The situation is completely reversed (and therefore consistent) for establishments where the eco-innovation determined a decrease in employment.

Figure 13: Change in overall employment

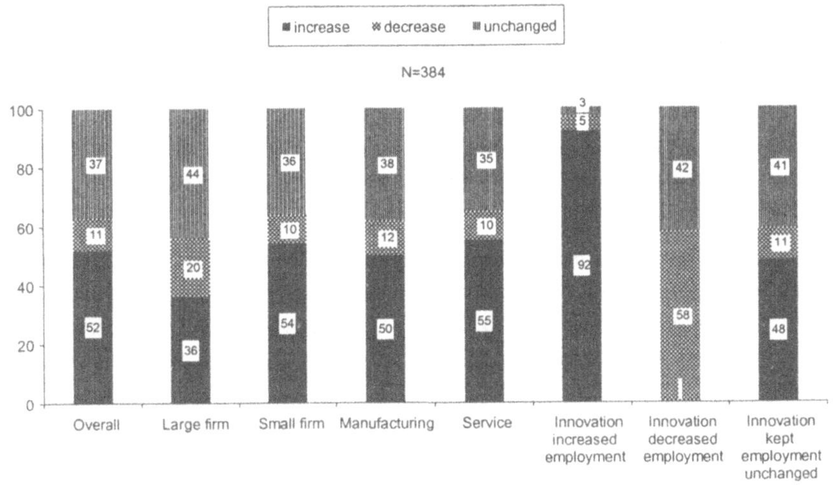

Figure 14 considers whether the establishment offers products or services that are specifically marketed as environmentally friendly and its relationship with employment effect. The results demonstrate that 77% of the companies that reported an increase in employment are also marketing their products with tangible environmental characteristics. Nevertheless it is difficult to draw conclusions since the number of companies which reported an employment change is too small.

Figure 14 : Does your company offer products or services that are specifically marketed as environmentally friendly

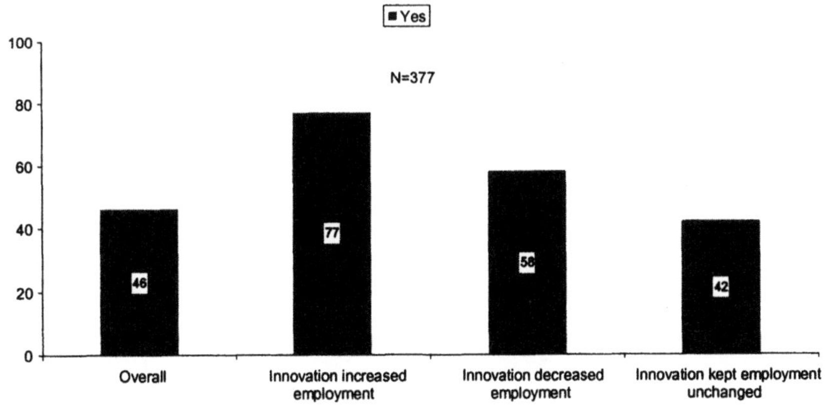

4 Concluding remarks

The Italian survey, conducted over a sample which is representative for the universe of Italian establishments, has shown that environmental innovation normally does not directly affect employment at establishment level. Nevertheless a positive correlation between environmental innovation and increase in employment exists, even if the size of this change is modest. The so called "soft" innovations, meaning those not requiring large investments on equipment and machinery, have a more positive employment effect.

A remarkable finding of the survey is related to the skill effect. Environmental innovations, regardless of the positive or negative employment effect, in nearly 80% of cases require some new skills or skill upgrade. The issue is dealt with training activities and outsourcing rather than via hiring permanent or temporary employees.

Annex: IMPRESS Questionnaire

For Interviewer:

Already known:

Industry/Manufacturing
Services
No information

Sector: ____

NACE Code: ____

Sectors included in IMPRESS survey are:

Industry/manufacturing:

Manufacturing	D
Electricity, gas and water	E
Construction	F

Services:

Wholesale, retail and trade	G
Hotels and restaurants	H
Transport, storage and communication	I
Financial intermediation	J
Real estate, renting and business activity	K

Good morning/ afternoon, my name is ______________ and I am calling from ________, an independent market research company.

Ask to speak to:

1. Head environmental department,
2. if 1 does not exist manager responsible for environmental issues,
3. if 2 does not exist head of establishment,
4. if 3 does not exist head of firm.

Contact with respondent

We are conducting an international survey funded by the European Commission on innovations that lead to environmental benefits – saving on the use of natural resources, generating less pollution, and the like. Innovation projects can intentionally seek environmental benefits, or such benefits can be unintended.

This survey is completely confidential.

Q1 Would you be willing to answer a few questions to see if this survey is relevant to your firm?

Yes	1	CONTINUE
No	2	GO TO Q2

ASK IF NO AT Q1

Q2 Is there a better time to call you back?

Yes	1	NOTE DOWN TIME AND DATE
No	2	ASK Q3

ASK IF NO AT Q2

Q3 Can you suggest someone else who could participate in this survey?

Yes	1	**NOTE DOWN NAME AND NUMBER** ______
No	2	**THANK AND CLOSE**

Categories of eco-innovations

Q4 Would you briefly describe your position within your firm?

Head of firm	yes/no
Head of establishment	yes/no
Head environmental department	yes/no
Environmental manager	yes/no
Health/Safety manager	yes/no

other (please specify)______________________________

Q5 Is this a multi-establishment firm?

Yes	[GO TO Q6]
No	[GO TO Q7]

Q6 Is this the head office?

Yes	[GO TO Q6A]
No	[GO TO Q6B]

Q6A Please base your answers on the establishment you are best informed about

Q6B Please base your answers on your establishment

Q7 In the last three years, did your establishment introduce any **pollution control** technologies?

- Yes ☐
- No ☐
- Dk (Don't know) ☐

Q8 In the last three years, did your establishment introduce any **recycling** programmes?

- Yes ☐
- No ☐
- Dk ☐

Q9 In the last three years, did your establishment introduce any new or improved **products or services** that are more **environmentally-friendly** than those already on the market?

- Yes ☐
- No ☐
- Dk ☐

Q10 In the last three years, did your establishment introduce any new or improved **processes** with environmental benefits?

- Yes ☐
- No ☐
- Dk ☐

Q11 In the last three years, did your establishment introduce any **organisational** innovations such as environmental reports, audits, or management programmes?

- Yes ☐
- No ☐
- Dk ☐

Q12 In the last three years, did your establishment introduce any new or improved **delivery, transport,** or **distribution** systems for its products or services, with environmental benefits?

Yes ☐

No ☐

Dk ☐

FILTER: IF RESPONDENT ANSWERS NO TO ALL CATEGORIES, END INTERVIEW

Otherwise read out:

Your answers show that your firm is eligible for this survey, which will take up to 20 minutes. If you agree, we will send you a copy of the study report. Can we proceed?

Concrete eco-innovation

I would now like you to think of the most **environmentally beneficial innovation** that your establishment introduced in the last three years. Environmental benefits include saving natural resources, generating less pollution, and the like. This can be a process, product, service, distribution system, organisational method, recycling system, or pollution control technology. The environmental benefits could have been intentional or a side-effect.

Q13 Would you please briefly describe this innovation and its environmental benefits?

- Note for interviewers: if the respondent cannot tell or if all were important, the person should make a decision based on what they are best informed about.

[if reluctant to describe it] You do not need to describe this innovation if you do not wish to, but it would help if you could answer the following questions about it in any case.

Q14 Which of the following categories describe this innovation:

	Yes	No
1. Product	☐	☐
2. Service	☐	☐
3. Distribution system	☐	☐
4. Process	☐	☐
5. Organisational method	☐	☐
6. Recycling system	☐	☐
7. Pollution control (end-of-pipe)	☐	☐

8. Other (if mentioned by respondent only): ________________________

Please keep this innovation in mind. The next set of questions only concern this innovation.

Q15 In which year was this innovation first introduced by your organisation?

YEAR______

Q16 Had this innovation been previously implemented by another firm or organisation?

Yes ☐

No ☐

Dk ☐

Q17 Who developed this innovation?

Your establishment or firm ☐

Other firms or organizations ☐

Your establishment or firm together with other firms or organizations ☐

Other________________ ☐

Q18 Did your establishment receive any government subsidies or grants to develop or purchase this innovation? *[Note: includes subsidies or grants from European Commission]*

Yes	☐
No	☐
Dk	☐

Q19 Approximately how much did your establishment invest in this innovation?

____

- Note for interviewers: If respondent says that innovation has not been financed by establishment but by entire firm, please repeat (READ OUT) Q6A or Q6B depending on answer to filter Q6. Firm data on innovation investments cannot be used for our analysis and is thus not relevant for us.

If the respondent cannot or will not answer:

Were your investment costs:

Less than 50,000 €	☐
Between 50,000 and 0.5 mln €	☐
Between 0.5 and 5 mln €	☐
More than 5 mln €	☐
Dk	☐

Q20 What percentage of your establishment's total innovation expenditures over the past three years were spent on this innovation?

Less than 5%	☐
Between 5% and 25%	☐
Between 25% and 50%	☐
Over 50%	☐
Dk	☐

Q21 Were any of the following factors **important** reasons for introducing this innovation?

Comply with environmental regulations	Yes ☐ No ☐ Dk ☐

Secure existing markets	Yes ☐ No ☐ Dk ☐
Increase market share	Yes ☐ No ☐ Dk ☐
Reduce costs	Yes ☐ No ☐ Dk ☐
Improve firm's image	Yes ☐ No ☐ Dk ☐
Respond to a competitor's innovation	Yes ☐ No ☐ Dk ☐
Achieve an accreditation	Yes ☐ No ☐ Dk ☐

Impacts on costs and sales

Q22 Did this innovation directly increase or decrease total unit sales?

Increase sales	☐	
Decrease sales	☐	
No effect	☐	[GO TO QUESTION Q23]
Dk	☐	[GO TO QUESTION Q23]

Would you estimate the percentage [increase/decrease] in sales:

Less than 5%	☐
5% to 25%	☐
Over 25%	☐
Do not know	☐

Read out Do not Know here!

Q23 Did this innovation lead to an increase or decrease in prices?

Increase prices ☐

Decrease prices ☐

No effect ☐ [GO TO QUESTION Q24]

Dk ☐ [GO TO QUESTION Q24]

Would you estimate the percentage [increase/decrease] in prices:

Less than 5% ☐

5% to 25% ☐

Over 25% ☐

Do not know ☐

Note for the interviewer: "Costs" are costs per unit.

Q24 Did this innovation increase or decrease your **energy costs?**

Increased energy costs ☐

Decreased energy costs ☐

No notable effect ☐ [GO TO QUESTION Q25]

Dk ☐ [GO TO QUESTION Q25]

Would you estimate the percentage [increase/decrease] in energy costs:

Less than 5% ☐

5% to 25% ☐

Over 25% ☐

Do not know ☐

Q25 Did this innovation increase or decrease your material costs?

Increased material costs ☐

Decreased material costs ☐

No notable effect ☐ [GO TO QUESTION Q26]

Dk ☐ [GO TO QUESTION Q26]

Would you estimate the percentage [increase/decrease] in material costs:

Less than 5% ☐

5% to 25% ☐

Over 25% ☐

Do not know ☐

Q26 Did this innovation increase or decrease your **waste disposal costs?**

Increased disposal costs ☐

Decreased disposal costs ☐

No notable effect ☐ [GO TO QUESTION Q27]

Dk ☐ [GO TO QUESTION Q27]

Would you estimate the percentage [increase/decrease] in disposal costs:

Less than 5% ☐

5% to 25% ☐

Over 25% ☐

Do not know ☐

Q27 Did this innovation increase or decrease your **labour** costs?

Increased labour costs ☐

Decreased labour costs ☐

No notable effect ☐ [*]

Dk ☐ [*]

*[GO TO QUESTION Q28 AND NOTE FILTER ABOVE Q28]

Would you estimate the percentage [increase/decrease] in labour costs:

Less than 5% ☐

5% to 25% ☐

Over 25% ☐

Do not know ☐

Effects on end-of-pipe measures, other products

[IF Q14.4 = YES AND Q14.1=NO AND Q14.2=NO, GO TO Q28, OTHERWISE GO TO Q30 AND NOTE FILER IN Q30]

Note: Attention to Coding!

Q28 Did the introduction of this process innovation replace, at least in part:

A previous production process in your firm?	Yes ☐	No ☐	Dk ☐
End-of-pipe pollution control equipment?	Yes ☐	No ☐	Dk ☐

Q29 Did this process innovation require:

Major reorganisation of your production system	Yes ☐	No ☐	Dk ☐
Substantial technical improvements	Yes ☐	No ☐	Dk ☐

Q30 [IF Q14.1 or Q14.2= YES, GO TO Q30, OTHERWISE GO TO Q32]

Did this innovation replace some of the [product/service] sales of your establishment?

Yes ☐

No ☐ [GO TO QUESTION Q32]

Q31 Can you estimate the percentage by which this innovation has replaced sales of the previous product/service? __________%

Employment effects of eco-innovation

Q32 How many man-years were required in **your establishment** for this innovation, from the original idea to its implementation?

Less than one	☐
One to ten	☐
Ten to 100	☐
Over 100	☐
Dk	☐

Q33 By long-term employment we mean employment in the firm for more than one year. Has this innovation increased, decreased, or had no noticable effect on the **number** of **longterm** employees in your establishment?

Increased	☐	[GO TO Q34]
Decreased	☐	[GO TO Q36]
No noticable effect	☐	[GO TO Q38]
Dk	☐	[GO TO Q38]

Q34 Please estimate the net number of **new** long-term employees, in full-time equivalents, due to this innovation __________

Q35 Did your establishment use any of the following methods to increase long-term employment?

Hire new employees	Yes ☐	No ☐	Dk ☐
Bring in employees from another location of your firm	Yes ☐	No ☐	Dk ☐

Q36 Please estimate the number of long-term employees, in full-time equivalents, **lost** due to this innovation __________

Q37 Did your establishment use any of the following methods to reduce long-term employment?

Redundancy, early retirement or natural wastage	Yes ☐	No ☐	Dk ☐
Move employees to another location of your firm	Yes ☐	No ☐	Dk ☐
Outsource specific tasks	Yes ☐	No ☐	Dk ☐

Skills

Q38 Did the development or regular use of this innovation require substantially new skills?

Yes

No [Go to Q43]

Dk [Go to Q43]

Q39 Did your establishment use any of the following methods to meet the need for new skills?

Obtain new employees from other units of your firm	Yes ☐	No ☐	Dk ☐
Train existing employees	Yes ☐	No ☐	Dk ☐
Hire new employees on a permanent basis	Yes ☐	No ☐	Dk ☐
Hire new employees on a temporary basis	Yes ☐	No ☐	Dk ☐
Outsourced work	Yes ☐	No ☐	Dk ☐

Other [IF MENTIONED BY RESPONDENT ONLY]: ____________________

Q40 [If YES TO TRAINING OPTION AND IF MANUFACTURING FIRM (NACE D,E,F)]

Which of the following types of employees received training because of this innovation?

Production workers	Yes ☐	No ☐	Dk ☐
Service workers	Yes ☐	No ☐	Dk ☐
Production engineering staff	Yes ☐	No ☐	Dk ☐
R&D staff	Yes ☐	No ☐	Dk ☐
Distribution and sales staff	Yes ☐	No ☐	Dk ☐
Management or other office staff not included above	Yes ☐	No ☐	Dk ☐

Other [IF MENTIONED BY RESPONDENT ONLY]: ____________________

Q41 [If YES TO TRAINING OPTION AND IF SERVICE FIRM (NACE CODE G,H,I,J,K)]

Which of the following types of employees received training because of this innovation?

Production workers	Yes ☐	No ☐	Dk ☐
Customer-facing service staff	Yes ☐	No ☐	Dk ☐
Production Engineering Staff	Yes ☐	No ☐	Dk ☐
R&D staff	Yes ☐	No ☐	Dk ☐
Distribution and sales staff	Yes ☐	No ☐	Dk ☐
Management or other office staff not included above	Yes ☐	No ☐	Dk ☐

Other [IF MENTIONED BY RESPONDENT ONLY]: ____________________

Q42 Did staff with the following qualifications receive training because of this innovation?

College or university degree	Yes ☐ No ☐ Dk/NA ☐
Skilled workers	Yes ☐ No ☐ Dk/NA ☐
Unskilled workers	Yes ☐ No ☐ Dk/NA ☐

Market, market structure, degree of competition

This brings us to the end of the questions on a specific innovation. I would now like to ask a few general questions about your establishment.

Q43 Does your establishment offer products or services that are specifically marketed as environmentally friendly?

Yes ☐

No ☐ [GO TO QUESTION Q44]

Dk ☐ [GO TO QUESTION Q44]

Are these:

Consumer products?	Yes ☐ No ☐ Dk ☐
Intermediate products sold to other firms or org.?	Yes ☐ No ☐ Dk ☐

Q44 Are each of the following an **important basis of competition** between your establishment and your competitors?

1. Price	Yes ☐ No ☐ Dk ☐
2. Quality	Yes ☐ No ☐ Dk ☐
3. Environmentally friendly features	Yes ☐ No ☐ Dk ☐
4. Innovative products or services	Yes ☐ No ☐ Dk ☐
5. Corporate image	Yes ☐ No ☐ Dk ☐

[IF MORE THAN ONE 'YES'] Which is the **most** important basis of competition?

1. Price	Yes ☐
2. Quality	Yes ☐
3. Environmentally friendly features	Yes ☐
4. Innovative products or services	Yes ☐
5. Corporate image	Yes ☐

Q45 Overall, how important are environmental regulations in determining the types of **processes** used by your establishment?

Very important	☐
Moderately important	☐
No effect	☐
Dk	☐

Q46 Overall, how important are environmental regulations in determining the types of **products** or services that your establishment can produce and sell?

Very important	☐
Moderately important	☐
No effect	☐
Dk	☐

Q47 In the last three years, has your establishment changed its processes or products **specifically** to comply with environmental regulations?

Yes	☐
No	☐
Dk	☐

General questions on firm

Q48 Approximately how many employees did your firm have in 1999?____

[IF RESPONDENT UNABLE TO GIVE A NUMBER, OTHER WISE GO TO Q49]

Would you be able to roughly estimate the total number of your employees, as follows:

Less than 50	☐
50 to 99	☐
100 to 249	☐
250 to 499	☐
Over 500	☐
Dk	☐

Q49 In the last three years, has total employment increased or decreased?

Increased	[GO TO Q49A]
Decreased	[GO TO Q49B]
No notable change	[GO TO Q50]

Q49A By what percent did employment increase? ____%

Q49B By what percent did employment decrease? ____%

Q50 Approximately what percentage of your establishment's **total** employees have a college or university degree? ______% Dk ☐

Q51 In the last three years, has the number of employees with a college or university degree increased by more than 5%, decreased by more than 5%, or remained stable?

Increased more than 5%	☐
Decreased more than 5%	☐
Stable	☐
Dk	☐

Q52 How many of your 1999 employees worked part-time? ____ Number

If unable to answer, then:

Can you estimate the percentage of workers employed part-time?______%

Q53 Approximately what was your establishment's total sales in 1999?____

[IF RESPONDENT UNABLE TO GIVE A NUMBER, OTHER WISE GO TO Q54]

Would you be able to roughly estimate your total sales, as follows:

- Less than 5 mln € ☐
- Between 5 and 25 mln € ☐
- 25 to 50 mln € ☐
- Over 50 mln € ☐
- Dk ☐

Q54 In the last three years, has your establishment's total sales:

- Increased ☐
- Decreased ☐
- No notable change ☐

If [increased/decreased]

Can you estimate the percentage [increase/decrease] in your total sales ____%

If they do not know:

- Less than 5% ☐
- 5% to 25% ☐
- Over 25% ☐
- Do not know ☐

Q55 In the next three years, do you expect your establishment's total sales to:

- Increase ☐
- Decrease ☐
- No notable change ☐

Q56 In 1999, labour costs accounted for what percentage of your total sales?___%

If they do not know:

- Less than 5% ☐
- 5% to 25% ☐
- 25% to 50% ☐
- 50% to 75% ☐
- Over 75% ☐
- Do not know ☐

This brings us to the end of the survey. Thank you for your appreciated cooperation.

Q57 To whom may we send the report? Please give me the name and the mailing address for the survey report.

Contributors

Anthony Arundel,

Maastricht Economic Research Institute on Innovation and Technology (MERIT), Maastricht (The Netherlands)

Matteo Bartolomeo,

Fondazione Eni Enrico Mattei (FEEM), Milano (Italy)

Mathias Binswanger,

Solothurn University of Applied Sciences (FHSO), Olten (Switzerland)

Rita Canu,

Fondazione Eni Enrico Mattei (FEEM), Milano (Italy)

Leann Chervenic-Poeth,

Maastricht Economic Research Institute on Innovation and Technology (MERIT), Maastricht (The Netherlands)

Ivo Demandt,

Maastricht Economic Research Institute on Innovation and Technology (MERIT), Maastricht (The Netherlands)

Steven Glynn,

Policy Research in Engineering, Science and Technology (PREST), Victoria University of Manchester (United Kingdom)

Najib Harabi,

Solothurn University of Applied Sciences (FHSO), Olten (Switzerland)

Annette Jochem,

Solothurn University of Applied Sciences (FHSO), Olten (Switzerland)

René Kemp,

Maastricht Economic Research Institute on Innovation and Technology (MERIT), Maastricht (The Netherlands)

Andrea Marsanich,

Fondazione Eni Enrico Matteo (FEEM), Milano (Italy)

Suhita Osório-Peters,

Centre for European Economic Research (ZEW), Mannheim (Germany)

Neville Reeve,

Policy Research in Engineering, Science and Technology (PREST), Victoria University of Manchester (United Kingdom)

Klaus Rennings,

Centre for European Economic Research (ZEW), Mannheim (Germany)

Daniel Schulthess,

Solothurn University of Applied Sciences (FHSO), Olten (Switzerland)

Thomas Zwick,

Centre for European Economic Research (ZEW), Mannheim (Germany)

GPSR Compliance

The European Union's (EU) General Product Safety Regulation (GPSR) is a set of rules that requires consumer products to be safe and our obligations to ensure this.

If you have any concerns about our products, you can contact us on ProductSafety@springernature.com

In case Publisher is established outside the EU, the EU authorized representative is:

Springer Nature Customer Service Center GmbH
Europaplatz 3
69115 Heidelberg, Germany

Batch number: 10725973

Printed by Printforce, the Netherlands